SLEEP

SLEEP

*A Groundbreaking Guide to the Mysteries,
the Problems, and the Solutions*

CARLOS H. SCHENCK, M.D.

AVERY

a member of Penguin Group (USA) Inc.

New York

AVERY

Published by the Penguin Group

Penguin Group (USA) Inc., 375 Hudson Street, New York, New York 10014, USA •
Penguin Group (Canada), 90 Eglinton Avenue East, Suite 700, Toronto, Ontario
M4P 2Y3, Canada (a division of Pearson Penguin Canada Inc.) • Penguin Books Ltd,
80 Strand, London WC2R 0RL, England • Penguin Ireland, 25 St Stephen's Green,
Dublin 2, Ireland (a division of Penguin Books Ltd) • Penguin Group (Australia),
250 Camberwell Road, Camberwell, Victoria 3124, Australia (a division of Pearson Australia
Group Pty Ltd) • Penguin Books India Pvt Ltd, 11 Community Centre, Panchsheel Park,
New Delhi–110 017, India • Penguin Group (NZ), 67 Apollo Drive, Rosedale, North
Shore 0632, New Zealand (a division of Pearson New Zealand Ltd) • Penguin Books
(South Africa) (Pty) Ltd, 24 Sturdee Avenue, Rosebank, Johannesburg 2196, South Africa

Penguin Books Ltd, Registered Offices:
80 Strand, London WC2R 0RL, England

First trade paperback edition 2008
Copyright © 2007 by Extreme-Nights, LLC

The Library of Congress catalogued the hardcover as follows:

Schenck, Carlos.
Sleep : a groundbreaking guide to the mysteries, the problems, and the solutions /
Carlos H. Schenck.
p.   cm.
Includes bibliographical references and index.
ISBN 978-1-58333-270-2
1. Sleep disorders.   I. Title.
RC547.S34   2007   200603720
616.8'498—dc22

(Paperback edition) ISBN-13: 978-1-58333-301-3; ISBN-10: 1-58333-301-0

Printed in the United States of America
1   3   5   7   9   10   8   6   4   2

BOOK DESIGN BY AMANDA DEWEY

Acknowledgments

This book was made possible by a special collaborative effort. Jessica Lichtenstein from the Joelle Delbourgo Agency first contacted me and later helped prepare the book proposal, with Joelle Delbourgo serving as my astute book agent. Megan Newman, the head of Avery, immediately grasped the broad and timely relevance of this book. Jenna Glatzer assisted me with care and skill in writing the book. Lucia Watson was my meticulous and perceptive editor at Avery, with earlier input from Dara Stewart. My patients who shared their stories, and also a poem, deserve my lasting appreciation. The members of the Minnesota Regional Sleep Disorders Center, with outstanding sleep technologists, nurses, physicians, and support staff, and with Mark W. Mahowald, M.D., as its director and my indispensable collaborator, deserve the highest tribute. My chief of psychiatry at Hennepin County Medical Center and Hennepin Faculty Associates in Minneapolis, Michael K. Popkin, M.D., is gratefully acknowledged for his support. The rapidly growing numbers (on an international scale) of sleep, parasomnia, and neuroscience researchers—my colleagues—are enthusiastically saluted. Their insights, discoveries, and therapies permeate this book. Finally, my wife, Andrea, who shared her own sleep stories, has given me loving support and encouragement through every step of the way.

CONTENTS

PREFACE

I have wanted to be a doctor since childhood, thanks in part to my pediatrician, a serious and caring man who always seemed to know what was wrong and how to fix it whenever I became ill. I received my medical degree in 1976 from the State University of New York at Buffalo School of Medicine, and after my internship I went to the University of Minnesota for further training. "Sleep disorders" wasn't on my radar screen as a specialty I wanted to practice at the time, since it was barely on any radar screen. The field of sleep medicine was in its infancy, and medical school curricula across the country spent little or no time on sleep and its disorders. I was fortunate, however, to witness a memorable lecture on sleep as an active state of the brain in my neurophysiology course as a first-year medical student. My professor explained that the brain is very active and communicating with itself during sleep. This seemed quite mysterious to me, and it was my introduction to the fascinating concept of *sleep neurophysiology,* that is, the functioning of the brain and the entire nervous system during sleep.

In 1982, as a staff psychiatrist at the Hennepin County Medical Center and faculty member at the University of Minnesota Medical School in Minneapolis, I was asked to join the recently formed Minnesota Regional Sleep Disorders Center. This initiated my career in sleep medicine, which to me felt like a natural fit, as I was able to apply knowledge on the clinical neurophysiology of sleep with my medical training to help diagnose and treat patients with a broad range of sleep problems.

I remember vividly the first patient I evaluated at the sleep center. He was an older man with an insomnia complaint who had early-morning awakenings (at 4 a.m.) with melancholic mood associated with a severe depression that had persisted for several months. The insomnia and depression completely disappeared with the antidepressant medication therapy I had proposed. My first patient, then, had a typical sleep problem linked with a common psychiatric disorder, and standard treatment brought about the expected outcome. *Things make sense in the sleep clinic,* I thought.

The second patient I saw was Mr. Donald Dorff, a 67-year-old recently retired grocery-store manager from Golden Valley, Minnesota, who complained of acting out "violent moving nightmares"—an unprecedented clinical complaint. Don had never been a sleepwalker, nor had he been subjected to night terrors. His regular doctor and a psychiatrist had found nothing wrong with him, so they decided to send Don to a sleep center— a novel type of medical referral at the time.

Don was a pleasant, affable man who enjoyed his large family and was an avid golfer all year round in Minnesota and in Florida. What brought Don to see me was that he had recently gashed his forehead from running straight into a chest of drawers while dreaming he was a football player running with the ball through the line of scrimmage. In fact, he was regularly acting out dreams and charging out of bed during the night. Don's dream plots mostly involved confrontations and fights with unfamiliar people and animals. In these nightmares, Don was placed in threatening situations and had to fight or escape in order to survive. Rather like B-movie-thriller plots, Don's dreams were full of action and lacking in character—classic fight-or-flight scenarios.

After we studied Don overnight in our center's sleep lab, we determined that Don had abnormal behaviors during sleep that were completely restricted to the rapid eye movement (REM) stage. This wasn't in the books. Something very different was going on with Don, since no one is supposed to move even a finger during REM sleep.

We realized that Don Dorff had a new type of sleep disorder. Don did not have spells of sleepwalking, night terrors, nocturnal epileptic seizures, acid reflux, sleep apnea, or other problems that could have possibly triggered his violent moving nightmares. The reality was that Don was demonstrating strange and harmful behaviors during his REM sleep in association with action-packed and confrontational dreams. He did not have the normal muscular paralysis of REM sleep that humans and other mammals are programmed to have in order to prevent the simultaneous acting-out of our dreams. Don was a danger to himself during his dreams.

And so my second patient at the sleep center jump-started my career in sleep medicine. He was our first patient with REM sleep behavior disorder (RBD), which we formally identified in the journal *Sleep* in 1986[1] (with Don being the first in our initial series of five patients). We named the disorder in 1987 in a report on our first ten patients with RBD published in the *Journal of the American Medical Association*.[2]

I had to learn about the treatment of RBD by the seat of my pants—no one had heard of it before, and no one knew how to treat it, either. I embarked on a logical approach to treatment by prescribing REM sleep-suppressing tricyclic antidepressant medications, which I thought would reduce the amount of REM sleep that Don would have and thereby lessen the duration of his RBD episodes. However, logic failed: There was no benefit from these tricyclic medications, and they caused troublesome side effects.

Then I considered how Don had also shown "periodic limb movements" during his non-REM sleep—with leg jerks every 15 to 30 seconds. Reports at the time indicated that this condition could be treated with clonazepam (brand name Klonopin), a benzodiazepine anticonvulsant. So I then had Don try clonazepam as treatment of his RBD, and, lo and behold, his RBD went away along with the disturbing dreams. What a great development! Don was thrilled and I was relieved because his clinical disorder was now under control. I was also intrigued by how his disturbing dreams were controlled in tandem with control of his dream-enacting behaviors.

After diagnosing Don, I realized that I was embarking on a career while simultaneously crossing a threshold into a new frontier in sleep medicine. RBD is now recognized as one of the most clinically fascinating and scientifically important discoveries in sleep medicine. It occupies a prominent position in a busy crossroads of the neurosciences, sleep medicine, and many other scientific fields. Our original 1986 publication on RBD in *Sleep* became the fourth-most-frequently cited article in that journal's twenty-five-year history, and in 2002, the editor of *Sleep* requested that I write a review article on RBD as part of the journal's Silver Anniversary celebration.[3]

Back in the mid to late 1980s, while continuing to see more patients with RBD, I was also starting to see a growing number of patients who complained of sleep-eating, violent sleep terrors, troublesome sleepwalking (including sleep-driving), nocturnal multiple personality, and a broad range of other aberrant behaviors during sleep. I realized at this time that it was the group of disorders called *parasomnias* that intrigued me the most in the practice of sleep medicine. Parasomnias are abnormal behaviors and experiences that accompany sleep, and they can be very abnormal indeed. RBD is just one of many parasomnias that can emerge from the far side of sleep. Over time I have discovered that parasomnias can emerge in tandem with almost all other sleep disorders and their therapies, with obstructive sleep apnea being a prime example. In this way, parasomnias can serve as prisms for viewing the entire realm of sleep, its disorders, and its therapies.

In my 25 years of practicing sleep medicine and conducting sleep research, the work I have collaborated on with my colleagues at the Minnesota Regional Sleep Disorders Center has resulted in discovering, naming, and often successfully treating a broad range of parasomnias—the extreme sleep disorders. When I became a sleep doctor, every sleep professional had already heard of insomnia and sleep apnea, but not many had encountered people who developed new identities when they arose from sleep (one of our patients thought he was a jungle cat), or raided the refrigerator every night without even realizing it, or regularly acted out violent dreams at night (RBD). This new knowledge about the parasomnias has helped bring tightly together the fields of neurologic disorders and sleep disorders. We now know, for example, that RBD is a strong predictor of Parkinson's disease and related disorders in men 50 years of age and older. What is going on inside the brain to account for this link? Another mystery of sleep to be solved.

Our center's parasomnia research also includes the first systematic studies on sleep-related injury,[4] and sleep-related eating disorders.[5] We have been at the vanguard of the medical-legal aspects of the parasomnias[6] and have documented the entity of *parasomnia pseudo-suicide*—that is, cases where people were believed to have committed suicide but were

actually in the middle of a sleep-disorder episode and not consciously doing anything dangerous.[7]

Perhaps even more important for the average person suffering from a
sleep disorder, we have identified a broad range of abnormal sleep behaviors associated with more common problems such as obstructive sleep
apnea, insomnia, and restless legs syndrome. (Obstructive sleep apnea is by
far the most common reason patients are evaluated overnight in the sleep
lab.) We have come up with successful treatments for the disabling condition called *severe morning sleep inertia*,[8] in which the afflicted cannot awaken
in a timely manner for work or school, despite heroic measures taken. And
not only will this book show how the links between common sleep problems and extreme sleep disorders affect the sleeping person, but it will address how they affect spouses and family members as well.

Because of our extensive research, the Executive Board of the American Academy of Sleep Medicine asked Mark W. Mahowald, M.D., the
sleep center director, and me to become the chairmen of the Parasomnias
Committee when they decided to revise the diagnostic manual of sleep
medicine, the *International Classification of Sleep Disorders* (ICSD), published
in 2005 as the ICSD-2.

Every Tuesday, the sleep center's conference room functions as a nerve
center at our hospital. The lunch-hour meeting pulls together physicians
from a wide range of disciplines, along with our stellar sleep technologists
and nurses, and we all compare clinical notes and discuss interesting and
difficult cases and the sleep-laboratory findings. This multidisciplinary meeting fosters cross-fertilization, helping us gain an even greater understanding of the links between our respective fields, and it has benefited the care
of our patients with a broad range of sleep problems.

Over the years, the print and television media have been captivated
by our parasomnia discoveries and have tapped into the many fascinating
stories shared by our patients and their families—and by the sleep-lab
video footage of patients "in action": eating, drinking, fighting, shouting,
singing, and engaging in all sorts of simple or extreme behaviors during
sleep and during mixed states of sleep and wakefulness.

I realized the time had come for a book for the general public on extreme
sleep disorders and how they relate to common sleep disorders. I have
never been on the speaker's bureau or advisory board of a pharmaceutical

company, nor have I ever been paid to promote a specific medication. Therefore, I have no conflict of interest when describing the therapies of sleep disorders in this book.

I've seen far too many people with extreme sleep disorders who remain "in the closet" and are reluctant to come out for fear of being mislabeled as having psychological problems, or out of concern that they will be accused of exaggerating what they do at night. I've had patients who would rather tie themselves to the bed every night than tell a doctor that they're having violent episodes or sexual behaviors during sleep. Many do not realize that effective treatments are available.

I hope this book will give real information and solutions to those who battle sleep disorders and offer a glimpse into the exciting world of sleep research and the many mysteries surrounding sleep that continue to be unearthed.

WHEN GOOD SLEEP
GOES BAD

I magine losing your children because you can't get out of bed. You have such trouble forcing yourself up in the morning that you set four alarm clocks, yet you still sleep right through them. When someone does finally wake you, through sheer perseverance, you fly into a rage and do your best to sneak back into bed because you're still exhausted. You can't keep a job because you consistently oversleep. You don't hear your baby crying, even though people always assured you that every mother instinctually tunes in to the sound of her baby's cries.

It's gotten so bad that you have to lock your children in their rooms so they don't harm themselves while you're sleeping. And what's worse, because you can't seem to get out of bed before noon, your children are essentially imprisoned for 16 hours a day—and you live with the terror that Child Protective Services could rightfully take them away from you at any moment.

That was my patient, Sarah, who came to me after a friend performed an important "intervention" and told her that something needed to change, *now*. Sarah wasn't just lazy or uncaring, despite what most people thought. She was suffering from a sleep disorder known as *severe morning sleep inertia,* a very real condition that can wreak havoc on a person's life.

In fact, there are many sleep disorders that can severely interfere with a person's daily life, and some that can be life-threatening to the sufferers or those around them. Severe morning sleep inertia and related sleep disorders are known as *parasomnias*—a word derived from the Greek prefix *para* (meaning "alongside of") combined with the Latin noun *somnus* (meaning "sleep") to account for the dictionary definition of "events that accompany sleep."

Sleep doctors define parasomnias as undesirable physical or emotional events that accompany sleep, occurring during entry into sleep, within sleep, or during arousals from sleep. The entire field of sleep medicine is fairly new, but the exploration of these parasomnias ("extreme" sleep disorders) is even newer. While people have surely always suffered from these disorders, we are just beginning to understand, classify, and treat them but the progress we have made in the past decade is astounding.

There are 12 core categories of parasomnias currently identified in ICSD-2. The *International Classification of Sleep Disorder* (ICSD), first published in 1990, has been a bible for sleep doctors. We've learned that parasomnias are much more common and occur across a broader age range than we ever knew. We've always known that children often sleepwalk and have sleep terrors, but we didn't know the prevalence of parasomnias in adulthood. Among Americans aged 15 to 65 years, at least 4 percent sleepwalk, at least 2 percent have sleep terrors, 10 percent have restless legs syndrome, at least 1 percent have sleep-related eating disorder, and 2 percent have sleep-related violence. To put that in perspective, we're talking about 8 *million* people in the United States alone who become violent in their sleep, and at least 16 million who sleepwalk.[1]

Unlike what many people would assume, a parasomnia—even a violent sleep disorder—typically does not signify that a person has an underlying psychiatric condition while awake. Those who exhibit violence during sleep, or scream and swear, or masturbate, or eat frozen ravioli, or wander into the hallway in their underwear while asleep, generally have no more of a psychological disorder than those who sleep peacefully every night. The man with RBD who tries to strangle his wife in the middle of the night probably *isn't* expressing any pent-up, suppressed rage. The woman who eats a whole box of chocolates while asleep probably *doesn't* have a daytime eating disorder. In most cases, the problem is strictly a faulty sleep mechanism.

Because of remarkable advances in diagnosis and treatment, most patients with a parasomnia can be treated successfully with medications, behavioral changes, or both. One of the great thrills for me as a sleep doctor is the knowledge that I can greatly improve the quality of life for most people who come to see me, even if they've had an extreme sleep disorder for decades.

Sarah's life transformed literally overnight once she was diagnosed and treated. She was able to be an attentive mother who awoke with her kids. I've seen these kinds of radical transformations over and over among people who thought they were beyond help. You're about to read some of their stories—from a woman who was binge-eating in her sleep to a man who acted out his dreams so violently that he often left his wife with bruises and black eyes. It is my hope that this book will lead to a greater understanding of parasomnias, and the broad range of other sleep disorders that can intersect with the parasomnias, and that if you see yourself or your loved ones in these stories, you'll also find resources here to help you and your loved ones sleep soundly. This book is also everyone's chance to take a peek and find out what happens in the mysterious, wild world of "extreme sleep."

SLEEP

Chapter 1

A TOUR OF SLEEP

Considering that people on average will spend 25 years of their lives asleep, it's surprising how little most of us know about what goes on when the lights go off. It's almost as if our waking selves and sleeping selves are two separate beings living in alternate dimensions, never catching more than a passing glimpse of each other.

Most people don't seem very curious or concerned about this odd desire they have to close their eyes, lie down, and blank out for several hours every night . . . until something goes wrong, or even very, very wrong. Maybe it's that they can't fall asleep, or can't wake up, or find chocolate syrup and raw noodles in their hair in the morning, or have just pummeled their spouse with jackhammer legs, and wonder what in the world happened when they thought they were in bed and supposedly asleep.

Poets and philosophers have painted contradictory images of sleep, but its resemblance to death is often a theme. Edgar Allan Poe called sleep "those little slices of death," John Fletcher named sleep "Brother to Death," and Roman poet Ovid wrote, "What else is sleep but chill death's likeness?" But maybe it's William Shakespeare, who often wrote about insomnia, who captured it best when he called sleep "nature's soft nurse."

While it may look like nothing much is happening while a person is sleeping, there's actually a complicated chain of events going on in the brain, and that chain is vital to our overall health. With all the possibilities for the complicated chain of events to break down or even go haywire, it's surprising that sleep clinics aren't exploding with more people reporting all sorts of sleep problems.

We know that at least 10 percent of the U.S. population has a clinically significant sleep disorder, but it's hard for us to know just how many people have parasomnias—because it's still such a new field of study. Most medical schools continue to allocate little time in their curricula for teaching about sleep and its disorders, so if someone goes to a family doctor with a sleep-related complaint that's out of the ordinary, it's unlikely to be diagnosed and treated properly. And worse, plenty of people don't realize that they have "real" disorders; they assume that their strange sleep behaviors are just their own weird quirks that they have to live with. Many of my own patients had no idea anyone else had ever gone through what they'd gone through until they saw a magazine article or television report where a patient from our clinic described the same problems.

There's also reluctance among many to talk about their concerns because of the potential stigma involved. Until recently, even the medical community believed that people who exhibited violent or aggressive or sexual behavior in their sleep probably had unaddressed psychological problems. Surely that man who throws punches and shouts obscenities in his sleep really has some pent-up rage he's not addressing during the day, right? As it turns out, probably not. There are rare cases of parasomnias that are caused by purely psychological issues, but we've now discovered that the vast majority of those with extreme sleep disorders have psychological profiles that look no different from the rest of the population. Their brain chemistry and behavior have gone amok during sleep.

To understand what makes things go so wrong during sleep, though, first we need to know what normal sleep looks like. So let's take a tour.

THE STAGES OF SLEEP

There are two basic types of sleep: *non-rapid eye movement* (NREM) and *rapid eye movement* (REM). NREM sleep has four stages, and REM sleep is one stage, so together, they comprise five important stages of sleep that cycle throughout the night.

Stage 1 (NREM)

This is the hazy stage when you're just falling asleep. It's a very light sleep, and it's easily disrupted—if someone turns on a light or makes a noise, you'll probably awaken. Often, if you're awakened from your first stage 1 sleep, you don't even know you were asleep. Or you may remember the experience as an interruption of a disjointed train of thought or image. Your *alpha rhythm* (the predominant rhythm—8 to 12 cycles per second— seen with eyes closed during relaxed wakefulness) is reduced to less than 50 percent of total brain-wave activity, your muscle tone becomes relaxed, your eyes slowly roll back and forth, and you may experience sudden muscle contractions *(hypnic jerks)* that jolt you awake, sometimes with the sensation that you were falling. We aren't sure yet why these contractions happen, but it's nothing to worry about. Some people, however, move rapidly or even race through stage 1 sleep and virtually jump into stages 2, 3, and 4 sleep.

Your brain activity, though mixed, is dominated by theta waves, which run at 4 to 7 cycles per second. This stage usually lasts less than 10 minutes, but can last 30 minutes or even two hours in people with sleep-onset difficulties.

To successfully enter sleep, a person must be able to disengage from wakefulness while being able to engage in sleep, a necessary two-step, mind-brain-body process with inherent vulnerabilities.

Stage 2 (NREM)

This stage is generally considered the "baseline" of sleep—the clear-cut starting point. It's difficult to pinpoint the exact second someone falls asleep because it doesn't happen all at once, and a person may report being awake when the tests appear to show stage 1 sleep, or alternating between wakefulness and stage 1 sleep.

In stage 2, you're still in a moderately light sleep, and still fairly easy to arouse, but your heart rate slows down and your temperature drops in preparation for the deeper sleep to come. Your brain activity, which can be measured by scalp electrodes in a sleep lab, will show slower waves mixed with sporadic bursts of faster waves called *sleep spindles* and also a brain wave pattern known as *K-complexes* that look like the letter "K". Sleep spindles and K-complexes are hallmarks of stage 2 sleep. As an adult, you'll spend about half of your total sleep time in stage 2.

Stage 3 (NREM)

Now you transition down to *slow-wave sleep* or *delta sleep.* Between 20 and 50 percent of this stage is made up of high-voltage, slow-wave activity known as *delta waves,* which run at 1 to 3 cycles per second. Stages 3 and 4 are often grouped together because there's not much that distinguishes them from a physiological standpoint.

Stage 4 (NREM)

It's a common misconception that REM sleep is the deepest sleep. Actually, stage 4 NREM is the deepest sleep, closely followed by stage 3. When you've been sleep deprived, your body craves delta sleep and will try to make up for lost time by getting you to stage 4 faster and keeping you there longer. When people say that someone is "out like a light" or in a "dead sleep," the person is probably in stage 3 or 4, when it's very difficult to arouse someone. This is the stage from which sleep terrors and sleepwalking are most likely to occur. Stage 4 is also the stage, it's thought, when the body does most of its repair work.

REM Sleep

We name this stage after the most easily seen physical marker: rapid eye movements. When a person is in the REM stage, he or she will often show bursts of eye flutters and back-and-forth movements, possibly in conjunction with dreams. This is the stage when the most vivid and prolonged dreaming occurs. Your brain is highly active (on sleep graphs, it looks almost the same as your brain activity when you're awake), but your body is temporarily paralyzed by an active, generalized inhibition originating in the lower brain. Your muscles—even the ancillary respiratory muscles other than the diaphragm—are "shut down" and completely inert, and it's believed that this is to protect us from acting out our dreams. Your heart rate picks up; blood flow to the brain increases; respiration becomes erratic, faster, and shallower, with increased oxygen consumption;

metabolism increases; sexual arousal occurs, with men developing erections and women clitoral engorgement; and blood pressure rises. Your brain activity shows low-voltage, fast-frequency waves. You also lose some of your ability to regulate your body temperature during this stage, so you're more susceptible to the temperature of your surroundings.

REM sleep is thought to be especially important in learning and memory processing, but there is much debate still about what exactly takes place. The ease of arousing someone from REM sleep varies, but it's usually far easier to wake someone from REM sleep than from stage 3 or 4 sleep. You can also become alert much more quickly when something wakes you out of REM sleep than when you are aroused out of delta NREM sleep. About 20 to 25 percent of an adult's total sleep time is spent in the REM stage.

THE CYCLE

You might think that the sleep cycle would progress as stages 1, 2, 3, and 4, then REM, then repeat, but it often doesn't occur in quite such a regular manner. Instead, a typical adult sleep cycle looks more like the following.

Stage 1 for up to 10 minutes, stage 2 for 10 to 25 minutes, stage 3 for about 5 minutes, stage 4 for 20 to 40 minutes, then back to stage 3 for just a minute or two, stage 2 for 5 to 10 minutes, *then* REM for up to 5 minutes.[1]

The first REM stage begins about 90 minutes into sleep, then the cycle starts again about every 90 minutes throughout the night.

During the earlier cycles, stages 3 and 4 are at their longest, but they get shorter and may drop out altogether in later cycles, with stage 2 lengthening to fill in the gap. The REM stage gets longer in the later cycles. In total, you'll spend about 75 to 80 percent of the night in NREM sleep and 20 to 25 percent in REM sleep.

The progression of sleep cycling appears to satisfy metabolic needs in the first part of the cycle (with increased amounts of delta NREM sleep) and brain activation in the second part of the cycle (with increased amounts of REM sleep).

It's normal to have brief awakenings during the night, particularly during the later cycles, around the transitions to and from REM sleep. With any luck, you won't even remember these awakenings, because they may last only a few seconds.

CIRCADIAN RHYTHMS

In addition to the importance of proper sleep cycling once we are asleep, there are other rhythms that play a key role in the timing of our sleep—the circadian rhythms.

The earth completes a rotation on its axis once every 24 hours, and as humans, we are built to respond to that rotation, scheduling our daily needs and tasks at appropriate times according to our environment, so that we are awake when there is sunlight and we are asleep after the sun has set and there is darkness. *Circadian* gets its meaning from the Latin roots *circa,* meaning "about," and *dies,* meaning "day." So a circadian rhythm is the cycle of "about a day." It's governed by your suprachiasmatic nucleus (SCN), which resides in the hypothalamus region of the brain, situated in the area between and just behind the eyes. Commonly referred to as an internal or biological clock, the SCN rules your body's natural programming, and it's adapted *(entrained)* to the approximate 24-hour day-night cycle.

The SCN picks up cues from the retina at the back of our eyes to determine when it's daytime and when it's night, judging by the amount of daylight or darkness. The SCN then passes messages around to other parts of the brain and body to direct the appropriate schedule for producing hormones, eating, and sleeping, and other vital functions. It determines when the hormone melatonin should be released from the pineal gland in the center of the brain—turned on at night and shut down during the day—to help us sleep at appropriate times.

People become keenly aware of their circadian rhythms when they attempt to change their sleeping patterns rapidly. For example, if you've taken a trip to a different time zone, you've undoubtedly experienced jet lag. This is because resetting the internal clock isn't a fast process. It can take several days (or even weeks) for the body to get adjusted to the new cues for day and night, and during that time, you'll probably feel sleepy and "foggy," and maybe irritable. Night-shift workers, or those whose work shifts change from week to week or from day to day, can also butt heads with their internal clocks.

The body doesn't want us to just pick any eight hours for sleep; it wants us to sleep in tune with the proper cues. We're made to feel alert when the daylight hits and to get sleepy when it's dark outside. Although you can psychologically get used to a schedule that falls outside of these parameters, you'll still be fighting your biological clock and be at risk for various health problems, which over the long run could include being at increased risk for depression, heart disease, and metabolic disorders.

We don't yet fully understand the interplay of external cues (light and dark) and our internal clock that creates the circadian pattern. Interestingly, without any external cues our internal circadian rhythm, driven by the SCN, falls out of sync with the normal 24-hour sun cycle, and on average becomes 24.2 hours.

AGE-RELATED CHANGES IN SLEEP

As you age, your sleep patterns age too.

Early Childhood

Newborns typically enter sleep the opposite way adults do; they go straight to REM sleep, known as *active sleep* in newborns. And their cycles are shorter than adult cycles: 50 to 60 minutes on average.

Starting around age one, babies shift to entering sleep through NREM stage 1, as they'll do for the rest of their lives. Young children have longer durations of slow-wave sleep (stages 3 and 4) than do adults, and it's even harder to arouse a young child from these stages of sleep as compared to an adult. Children are known to be prone to sleepwalking, and studies show that up to 17 percent of children sleepwalk, with the behavior peaking at age 11 or 12—but people rarely recognize how often sleepwalking continues into, or begins to occur in, adulthood. (4 percent of adults sleepwalk).

The National Sleep Foundation recommends 14 to 15 hours of sleep per 24 hours for infants (3 to 11 months), 12 to 14 hours for toddlers, 11 to 13 hours for preschoolers, and 10 to 11 hours for school-aged children, though they found in a 2004 poll that children consistently sleep less than the recommended amounts.[2] About 69 percent of children in these age groups have sleep problems at least a few times a week, according to the parents polled.

Adolescent and Young-Adult Years[3]

As children reach preteen and teenage years, their sleep recipes will change drastically; they will lose about 40 percent of their slow-wave sleep despite needing about the same amount of total sleep.

They also seem to be programmed in major conflict with most school schedules. During puberty, biological changes shift an adolescent's internal "clock," making it more natural for a preteen or teen to fall asleep around 11:00 p.m. to midnight. They still need about nine hours of sleep per night, the same as they did during their grade school years. However, because the school day may begin at 7:00 a.m. (or earlier, for those with before-school activities!), many adolescents don't get enough sleep, which results in daytime sleepiness and a tendency to fall asleep during class.

It can be very difficult for adolescents to fight their internal programming to stay alert at times when they feel like sleeping, or go to bed early at times when they feel wide awake. Building up a big sleep debt (from prolonged periods of sleep deprivation) can interfere with a person's ability to concentrate and learn, make the person feel more irritable, and have other adverse effects on health. The effects of chronic sleep deprivation can also be mistaken for attention deficit hyperactivity disorder (ADHD). Because of this, there have been efforts in some school districts to adopt later start times.

One of the more troubling effects of sleep deprivation is falling asleep behind the wheel: In a North Carolina study, more than half of all fall-asleep crashes involved people 25 years of age or younger.[4]

In August 2006 the impact of sleep deprivation again made tragic headlines when a Comair jet crashed in a Kentucky field because it took off from the wrong runway, killing 49 people. It was revealed that the sole air traffic controller on duty had slept for just two hours before his overnight shift. He had nine hours off in between work shifts, having worked all morning and part of the afternoon on Saturday, then returned Saturday night to work through Sunday morning. Investigators are studying his schedule to estimate how much of a role sleepiness may have played in this accident.

Early to Mid-Adulthood

During the early adult years, slow-wave sleep hasn't dropped much below its adolescent level, REM sleep remains constant, and you typically need less sleep than you did in earlier years.

The National Sleep Foundation polled adults between the ages of 18 and 54 and found that those in this age group often play a game of "catch-up," where they don't get enough sleep on weeknights (6.7 hours per night, on average), then try to compensate by sleeping more on weekend nights (7.6 hours per night, on average).[5] Work and school schedules clearly play a part in this pattern. About 16 percent of adults usually sleep less than 6 hours per night.

Different people have different sleep needs: Seven hours may be enough for one person to feel rested, whereas another may need more than nine hours to be at his best. Although most adults have heard that the magic number is eight hours, that's not true for everyone. Not all people who naturally sleep less than that are sleep deprived, and not all people who sleep more than that are lazy! Likewise, it appears that the tendency to be a "morning person" or a "night person" is something we're born with. We all have unique biological needs and do our best when we accept them instead of forcing ourselves to fight them. The important questions are how do we feel and how do we function during the daytime after getting a certain amount of sleep during most nights—if we feel good, then we are getting enough sleep, but if the sleep we're getting makes us feel tired, then either we need to get more sleep or else we have a sleep problem that needs to be addressed.

Later Adulthood

Studies have shown that up to half of women have sleep disturbances around the time they enter menopause. One study of menopausal women classified 42.1 percent of them as "poor sleepers," with depression and aging showing significant effects on quality of sleep.[6]

Elderly people have the most sleep difficulty of all, though, reporting high levels of insomnia, restless legs syndrome, sleep-related breathing disorders, and intrusive early-morning awakenings. Many factors can contribute to this, such as other health problems (painful conditions may keep a person awake) and medications. There's nothing about the aging process itself that should cause sleep problems.

In 2003, the National Sleep Foundation conducted a large-scale poll of Americans between the ages of 55 and 84 to learn more about their sleep behaviors in relation to their overall health, activities, moods, and outlooks on life.[7] They surveyed 1,506 people by phone and discovered something that should seem obvious but is rarely discussed: The better the person's overall health, the better his or her sleep. And vice versa: The higher the number of medical conditions, the more likely it is for the person to report sleep problems. This type of correlation also held true with lifestyle and outlooks: Those with more active lifestyles and a more positive outlook on life tended to have fewer sleep complaints.

On average, older people sleep 7 hours on weeknights and 7.1 hours on weekend nights, according to this poll. Sixty-seven percent report having sleep problems at least a few times a week, including difficulty falling asleep, frequent awakenings throughout the night, inability to go back to sleep after an unwanted awakening, pauses in breathing, and unpleasant feelings in their legs. Older women were more likely to report symptoms of insomnia, whereas older men were more likely to snore and have sleep-related breathing problems.

Only about 8 percent of respondents had actually been diagnosed with a sleep disorder, however.

WHY IS SLEEP IMPORTANT?

At a time when the Internet, Starbucks, and late-night television have turned us into a nation of the sleep-deprived and hyperalert, the concept of a "good night's sleep" seems elusive, maybe even a little old-fashioned. But the truth is that researchers have crossed a threshold into a new understanding of the importance of sleep.[8]

The question of why we sleep isn't as easy to answer as you might expect. Aristotle decided in 350 B.C. that it must have something to do with digestion. In his treatise *On Sleep and Sleeplessness,* translated by J. I. Beare, Aristotle explained that when food is digested, it releases hot vapors into the veins, which turn into blood and wind up in the heart. This makes a person feel heavy and lie down. The outer body parts get cool, which makes the eyelids droop. Then sleep occurs "until the purest part of this blood has been separated off into the upper parts of the body, and the most turbid into the lower parts. . . . Also, as a general rule, persons whose veins are inconspicuous, as well as those who are dwarf-like, or have abnormally large heads, are addicted to sleep," he explained.

Later, physicians decided that sleep must be the result of too much blood getting congested in the brain. Or maybe the opposite—too much blood had fallen *out* of the brain, making the brain shut down and the body lie horizontally to encourage the blood to get back in there.

Well, we've come a long way in our understanding of how sleep works, but we're still not sure exactly why it happens. Many experts believe it's to "recharge" our energy supplies and repair our bodies, much like charging up a cell phone after using it all day. Certain hormones, such as growth hormone, are released during sleep, and the body tissues go through repair processes.

Others think sleep's primary function is to process our memories, store them appropriately, and promote learning. It may be that sleep helps us to file away what we learned that day, and get rid of non-useful memories that are taking up room in the brain. Various other theories exist about the function of sleep, such as "sleep is of the brain and for the brain"—that is, optimizing brain function is the main purpose of sleep.

The best way to illustrate why we need sleep is to examine what happens when we *don't* get enough sleep during many or most nights—or when our sleep quality is poor:

- Concentration decreases.
- Decision-making ability declines.
- Irritability and frustration increase.
- Motor function is impaired.
- Speech is impaired.
- Car accidents are more common.
- Workplace accidents are more common.
- Ability to fight illness and disease declines.
- Mental and physical health disorders worsen.
- The aging process may be speeded up.

Sometimes the consequences of sleep deprivation are minor (like nodding off while watching a movie), but long-term sleep deprivation can have serious consequences like hallucinations and even suicidal thinking. Sometimes sleep deprivation can be catastrophic. Major disasters such as the nuclear power plant accident at Three Mile Island, the *Exxon Valdez* oil spill, the 1984 gas leak in Bhopal, and the disaster of the space shuttle *Challenger* were all officially attributed to sleepiness-related impaired judgment in the workplace. Major problems and disasters are much less likely to occur, and can be handled much more efficiently, when good sleepers are in charge.

Recent research from the Walter Reed Army Institute of Research has shown that soldiers who are substantially sleep deprived have diminished decision-making capabilities, constructive thinking, and "moral reasoning."[9, 10] Soldiers who were forced to stay awake for 53.5 hours tended to judge controversial solutions to moral dilemmas on a test as "acceptable" more often than they would when they were rested. In other words, there is diminished critical thinking in the sleep-deprived state.

So even if scientists can't agree on exactly why we sleep, we can all agree that it cycles through five predictable stages; it changes as we age; and it is essential for virtually all creatures, clearly serving multiple beneficial purposes. As we'll explore further in later chapters, we also know that parasomnias can emerge in males and females from any sleep stage at any time of life—even *in utero*.

Chapter 2

INSOMNIA

Bleary-eyed Nights

J ust about everyone knows the occasional feeling of lying in bed and staring at the clock, noting with misery how few hours are left until it's time to wake up the next morning. Most of us have a few favorite tricks for dealing with it, too—counting sheep, drinking warm milk, reading . . . but this problem can become quite resistant to our best attempts to overcome it. That's why insomnia ranks near the top of the reasons people seek medical attention.

Insomnia refers to repeated difficulty in falling asleep or staying asleep, waking up too early, and/or feeling unrefreshed or unrested the following day. If you had a hard time falling asleep, or seemed to wake up too much at night, but you functioned quite well the next day and didn't feel fatigued, you probably got an adequate amount of sleep.

Some people talk themselves into believing they have a sleep problem when they're really doing fine, because they've heard that "eight hours" is the magic number. But not everyone needs eight hours of sleep to feel refreshed. Some do fine on seven, or even six. In a different vein, some people want to believe that they can get away with pulling all-nighters, or that sleep really is expendable and can be stretched and tossed around. After all, geniuses throughout history reportedly slept only four or five hours a night, and sometimes had extremely idiosyncratic sleep schedules (Leonardo da

Vinci, for example). This type of public sentiment is where we get catch phrases like "Sleep is for wimps" and "I'll sleep when I'm dead." But the evidence is now clear: Sleep is not optional. Most of us cannot misuse or abuse our sleep and expect to get away with it—sleep deprivation can retaliate against us with a vengeance. In fact, obtaining a proper amount of sleep and good quality sleep should be considered top priorities in our lives.

In contrast, there are some people who were born with the innate (genetic) tendency for being "short sleepers." These people feel fine and function perfectly well during their waking hours despite getting only five or six hours of sleep every night. (It is extremely rare for people to acquire this tendency or to train themselves to be short sleepers.) People like Thomas Edison and Margaret Thatcher are often referred to as insomniacs, but that's probably not accurate—they were both short sleepers. Edison reportedly said that the need for sleep was overrated and that anyone could learn to go without it. (Edison did like to nap during the day, though, and was purportedly fired from an early job because his boss discovered that he had invented a device to make it look like he was working when he was actually asleep![1])

Those who have true insomnia, though, are usually well aware of it, and live unhappily with its direct consequences.

Insomnia, which affects about 58 percent of American adults at some point in their lives, is the most common of all sleep complaints and true sleep problems. Many people just see it as an annoyance rather than a serious health problem, but I urge you to take it seriously because it can have major and even catastrophic effects on your health, quality of life, mood, memory, and general performance. We have so many anti–drunk-driving campaigns, but where are the campaigns to get the insomniacs off the road? ("Friends don't let friends drive drowsy!" or "Don't be drowsy and drive.") Our sleep center's director, Mark W. Mahowald, M.D., has taken a major step at raising awareness by helping to create a new driver's education module that illustrates the dangers of driving while being sleepy.

It is likely that in excess of 100,000 motor vehicle crashes annually in the United States are due to driving while drowsy. Sleep deprivation has performance-degrading effects on the brain similar to alcohol, and can lead to accidents even if a person manages to stay awake. It can ruin a person's concentration and response time, and can make one increasingly vulnerable to the sedating effects of various medications, such as antihistamines.

One of my patients, Julie, shared with me a poem she wrote while experiencing insomnia.

Insomnia Series
II
By Julie J.

Who stole my nights?!?

plush, lush
warm slumber

i hear
the rest of you
snoring
     ever
        so
         gently

hazy, drifting
cobwebbed meanderings

taken for granted

dreaming boats on seas
i can never sail

you
are lost to me now
in deep undercurrents
restoring your sanity

while i
damned to be the waking dead

dry-docked, land-locked
marooned

> blank, bleak lightless light
> and late night TV
>
> herbal teas, drugs and this
> heart-rending attempt
>
> to CURSE YOU!
>
> Vampire of my sleep

DEFINING INSOMNIA

There are three types of insomnia:

1. **Transient:** Isolated cases of insomnia that last for a few weeks or less
2. **Intermittent:** Recurring transient insomnia
3. **Chronic:** Long-lasting insomnia occurring most nights for more than a month

Within these three types, insomnia can be categorized according to when it causes trouble.

Sleep-onset insomnia is the classification for when a person has trouble falling asleep in the first place. It's called *sleep-maintenance insomnia* when a person wakes up during the night and can't go back to sleep. Waking up too early in the morning is called simply *early-morning awakening*. A combination of any of these problems is known as *mixed insomnia.*

So, in other words, if Sally keeps waking up in the middle of the night for a couple of weeks (but then the problem stops), she had transient sleep maintenance insomnia. Brandon, who has trouble falling asleep on an ongoing basis for several months, has chronic sleep onset insomnia.

Insomnia is not defined by the total sleep time, and there's no "magic number" of minutes or hours a person must lie awake while lying still or tossing and turning before it's considered insomnia. On average, it takes people seven minutes to fall asleep. But you may need 20 or 30 minutes—

or even an hour—to drift off, and not consider it a problem. What matters far more than the amount of time it takes to fall asleep is how it affects you. If your functioning is impaired the following day, if you don't feel restored after sleep, or if you can't stand your lack of ability to fall asleep or stay asleep, it's time to take your insomnia seriously and seek help.

THE THREE FACES OF INSOMNIA

Even if the insomnia is fairly short-lived, it can still cause problems or wreak havoc. So let's see if we can shed some light on this common problem. There are three kinds of people who get insomnia, apart from those with medical or psychiatric disorders affecting their sleep. The first are people who do all the right things in preparing for sleep and who achieve a desired mellow feeling at the point of going to bed—but they still cannot cross the threshold to enter the realm of sleep.

The next are people who ordinarily have no trouble falling asleep but who electively stay awake until all hours of the night, either "burning their candles at both ends" to complete tasks, or engaging in various recreational activities, with the consequence being that their sleep is shortened during far too many nights. They have the mistaken belief that sleep is a ball that can be kicked around. Also, stress can interfere with our sleep.

The practical implication in starkly contrasting these two polar opposite versions of insomnia is this: We must look hard in the mirror when assessing our sleep, daytime habits, evening routines, and stress levels, and be thoroughly honest with ourselves about what we may be doing wrong to undermine our sleep, *but* on the other hand, we should not unnecessarily "guilt-trip" ourselves if we are doing nothing wrong and still cannot sleep.

Finally, there are people who have insomnia as an environmentally induced sleep disorder. Any sound or light disturbance in the bedroom or from the outside can adversely affect their sleep—including the bed partner's snoring, leg jerking or kicking, sleep-talking or sleepwalking, night terrors, sleep-eating, or violent dream enactment. In other words, a parasomnia in one bed partner can cause insomnia in the other bed partner. Yoked sleep disorders.

When I talk to patients about sleep-onset insomnia, I remind them that to fall asleep in a timely fashion, a person must be successful in disengaging from wakefulness *and* be successful in engaging in sleep. A person with difficulty falling asleep may have problems with either or both of these transition events.

WHO GETS INSOMNIA?

Anyone can get insomnia, but certain traits and situations put a person at higher risk.[2] Those include:

- **Advanced age.** The risk for insomnia increases with age; about one-third of people older than 65 have chronic insomnia.
- **Worriers.** Those who tend to have "worrier" personalities are more prone to have trouble relaxing and falling asleep.
- **Women.** In the 2005 Sleep in America Poll, conducted by the National Sleep Foundation, 57 percent of women reported that they frequently experience at least one symptom of insomnia, compared to 51 percent of men.
- **Psychiatric disorders.** Most psychiatric disorders, but particularly depression and anxiety disorders, make it more likely for a person to have insomnia.
- **Physical disorders.** Likewise, many medical conditions can disturb a person's sleep pattern and sleep depth, particularly painful disorders.
- **Alcohol use or abuse and drug abuse.** Even though some people turn to alcohol, thinking it will help them sleep, this actually harms the quality of their sleep and contributes to insomnia in the long run. Others who abuse alcohol or drugs often have major insomnia problems.
- **Certain jobs.** A fascinating study of 6,268 people in 40 different work groups compared the insomnia rates in each.[3] They discovered that 18.9 percent of male bus drivers complained of having "rather or very much" difficulty falling asleep, compared to 3.7 percent of male managers and 4.9 percent of male physicians. Other respondents reporting high symptoms of insomnia included female cleaners, male teachers, male and female laborers, male construction workers, and female hospital aides. Gardeners, social workers, and construction workers were most likely to use sleeping pills frequently.

The interpretations of these findings can be straightforward, as with painful disorders causing disrupted sleep, or can be puzzling, as with job-related insomnia. Clinicians should at least be aware of various occupational risks for insomnia, and clinical researchers should devise studies to tease out and identify the factors promoting insomnia within the at-risk jobs.

CAUSES OF INSOMNIA

Insomnia can have many causes, and chronic insomnia is often a multifactorial disorder. The causes of insomnia can be situational, physical, and psychological, or all three combined.

Situational Causes

- **Change in routine.** A change in your work or school schedule can disrupt your natural sleep rhythms.
- **Change in environment.** Sudden changes in the weather or sleeping in a strange place can cause insomnia.
- **Poor sleep conditions.** Noisy rooms; an uncomfortable bed; too much heat or cold; bad odors; a snoring, leg-jerking, or restless bed partner; flashing lights; or airplanes passing overhead can disrupt a normal night's sleep.

INJURY BY SLEEP DISRUPTION

In 2005, police in Japan arrested a 58-year-old woman and charged her with inflicting injury because she had deliberately made a hole in her back door so she could point a portable stereo at her neighbor's home and blast dance music almost around the clock for two years. Her victim, a 64-year-old, needed a month of medical treatment for insomnia and headaches.[4]

Physical Causes

- **General illness.** A cold, sore throat, or allergic congestion can keep you awake, or frequently interrupt your sleep.

- **General medical conditions.** Nearly all organ systems (heart disease, kidney disease, gastrointestinal problems, asthma, Alzheimer's disease, Parkinson's disease, hyperthyroidism, and so on) have been implicated in causing or worsening insomnia.
- **Pain.** Physical pain can prolong sleep onset or promote sleep disruption. Those who suffer from arthritis are particularly susceptible to insomnia.
- **Medications.** While some medications make you drowsy, others can have the opposite effect and act as stimulants, which if taken in the afternoon or evening can delay one's time for falling asleep. This is true of both prescription and over-the-counter medications. Therefore, starting new medication or changing the dosage of a medicine may be the culprit in your sleep disturbance.
- **Hormone fluctuations.** Women are more likely to suffer from insomnia during menstruation, pregnancy, menopause, and perimenopause with hot flashes.[5]
- **Caffeine.** Even if you drink coffee, tea, or caffeinated soda early in the day, it can still keep you up late at night, depending on how much you consume. Don't forget about the caffeine in chocolate, either.

Psychological Causes

- **Anxiety disorders.** Panic disorder, obsessive-compulsive disorder, generalized anxiety disorder, and post-traumatic stress disorder are all linked with insomnia.
- **Depression.** Depression is a particularly common cause of recurrent, unwanted early-morning awakenings. Depression can also be an *effect* of long-standing insomnia; insomnia puts patients at a much higher risk for major mood disorders, including the risk for suicide.
- **Stress.** Whether it's job-related, relationship-related, money-related, or from any other cause, stress is a major factor in insomnia. Even "good stress" (like an upcoming wedding or a friend coming to visit) can count as stress with adverse effects.
- **Worry about sleep.** It's a vicious cycle. Once you've experienced insomnia, you may worry about it the next night—which makes it difficult to sleep. Insomnia can become self-sustaining when you get anxious about its reappearance.

HYPERAROUSAL

There are also some people who have no apparent situational, psychological, or physical causes for their insomnia, and evidence now shows that some insomniacs may be in a constant state of hyperarousal. Many are actually *less* sleepy during the day than normal as measured by objective daytime nap studies, and they may also have an increase in metabolic rate across the 24-hour period. The interpretation of these findings goes back to the brain, which determines the set point for levels of drowsiness, arousal, and alertness. The earlier in life that insomnia manifests itself, including childhood, the more likely that there is a genetic basis for the presumed hyperaroused state responsible for the insomnia.

SLEEP IN THE CITY

Researcher Bert Sperling, known for his "Best Places" studies, conducted a study to determine where people were getting a good night's sleep, and where insomnia raged on in the United States. He analyzed information from the Centers for Disease Control and Prevention (CDC), the Bureau of Labor Statistics, and the U.S. Census Bureau. Here's a portion of his results.[6]

Best Cities for Sleep	*Worst Cities for Sleep*
1. Minneapolis, MN	1. Detroit, MI
2. Anaheim, CA	2. Cleveland, OH
3. San Diego, CA	3. Nashville, TN
4. Raleigh-Durham, NC	4. Cincinnati, OH
5. Washington, DC	5. New Orleans, LA

HELP FOR INSOMNIACS

There are many quick tips that may alleviate your sleeplessness.

1. **Don't just lie there.** If you've been trying to fall asleep and can't within a reasonable time, stop tormenting yourself. Get up. Walk

out of the room. Keep the lights low, but do something relaxing for a little while—read, hang out with a pet, listen to soft music, say prayers—and don't get back into bed until you're feeling sleepy again.

2. **Keep a consistent schedule.** One of the keys to getting a good night's sleep is to keep a fairly strict routine: Go to bed and wake up at the same time every day, as much as possible, even on weekends.

3. **Exercise early.** Exercise is great for your body and your mind, but do it during the day, not at night. Because exercise stimulates you, it can keep you up when you want to be down. (This may also apply to the timing of sexual activity.)

4. **Don't nap.** Try not to nap during the day; you want to actually be sleepy when your head hits the pillow at night.

5. **No multitasking in the bedroom.** You want to condition your brain that bedtime equals sleep time. Take the television out of there, don't bring work into the bedroom, fold laundry elsewhere . . . In short, keep the bedroom a special space for rest.

6. **Learn relaxation techniques.** Meditation, tai chi, yoga, guided imagery, progressive muscle relaxation, and similar techniques can help ease your stresses and "slow" your brain into a more relaxed state in general, which will make it easier to transition to sleep. Choose the method that is most suitable and works best for you.

7. **Make lists.** If you're having trouble sleeping because you're running through all the things you need to do the next day, write yourself a list of tomorrow's tasks. As you write each item, decide that you don't need to think about it anymore until tomorrow, because you've already written the reminder.

8. **Quit smoking.** Smokers often have nicotine cravings in the middle of the night, leading to unwanted arousals and difficulty falling back asleep. In addition, nicotine is a stimulant and can have bad effects on your nerves—as well as just about every function of your body. But you already knew that. Here's just one more reason to quit. If you are in the process of quitting smoking, consider putting on a nicotine patch at bedtime to minimize awakenings linked with cravings to smoke.

9. **No nightcaps.** Avoid alcohol close to bedtime. Even though it can help you get to sleep faster, it reduces the quality of your sleep, which can make you feel even worse the next day.

10. **Check your diet.** Avoid heavy meals and eating chocolate close to bedtime. A light snack (that doesn't involve a lot of sugar or caffeine!) may help you sleep, but a full meal can keep you awake while your body is busy digesting. Keep tabs on whether a protein-rich or a carbohydrate snack is of greater benefit to you for promoting sleepiness.

11. **Restrict your sleep.** Although it sounds counterintuitive at first, the idea here is that insomniacs spend too much time in bed and not enough time asleep; therefore, you want to recondition yourself to expect to sleep the entire time you're in bed. To do so, you may try restricting your time in bed. Stay out of the bedroom until, say, 3 a.m., and get up at the usual time in the morning. Every night, go to bed 15 minutes earlier, and still get up at the usual time, and so on until you stop where you want. This is called *sleep restriction/sleep consolidation therapy.* You'll truly be exhausted when you go to bed and fall asleep promptly. A person has to be quite motivated to successfully complete this therapy, but it is well worth the effort. This therapy may take a week or two, and the advisability of working and driving during this time period should be discussed with your doctor.

12. **Take a bath.** Try soaking in the tub before you go to bed. Lavender bath salt or oil is said to help people unwind.

13. **Massage.** Massage can be a great way to soothe stiff muscles and promote relaxation.

14. **Step away from the monitor.** Try to avoid television or computer activity just before bedtime. Staring into bright, flashing lights can trick your circadian clock into believing it's still daytime, and can throw off your sleep schedule.

15. **Reduce environmental distractions.** Use a sleep mask to block out unwanted light. Try earplugs, a white-noise machine, or a fan to mask street noise or sounds in the house. Experiment with the room's temperature to find the setting that feels most comfortable. Be creative in finding solutions! Every little bit helps.

OVER-THE-COUNTER MEDICATIONS

There is little published research supporting the use of these medications as therapy for insomnia. People who take over-the-counter drugs need to be cautious and be honest with themselves in assessing benefits and side effects. Over-the-counter sleep medicines are usually of limited help for insomniacs but can be very effective in some people. Most of these medications, such as Tylenol PM, Unisom, and Sominex, contain antihistamines, such as diphenhydramine, just like you'd use for a cold or allergies, and they could give you a "sleep hangover" the following morning. Do not use an over-the-counter sleep aid for longer than directed—preferably just a day or two. If insomnia becomes a more common problem, then consider seeing your physician before chronic insomnia sets in.

PRESCRIPTION MEDICATIONS

There are three classes of medications approved for the treatment of insomnia: benzodiazepines (such as temazepam, estazolam, and others); the newer, non-benzodiazepines (such as zolpidem [Ambien, Ambien CR], Zaleplon [Sonata], and eszopiclone [Lunesta]); and the melatonin receptor agonist ramelteon (Rozerem). Most of these agents will be helpful for most people who are being treated for insomnia, provided that the patients are properly diagnosed and have regular follow-ups. Also, some tailoring of therapy at times may be needed. Sonata, for example, has a one-hour duration of action, so it can be used for middle-of-the-night awakenings. Given the relatively level "therapeutic playing field" among the hypnotic medications, the cost to the patient should be considered when selecting a medication. Of course, the list of approved medications in the patient's health plan must be considered.

These medicines usually have a low incidence of side effects, and a low abuse potential. For example, our center published a study in the *American Journal of Medicine* in 1996 in which we analyzed various outcomes in 170 adult patients with long-standing sleep-disruptive disorders who were treated with nightly benzodiazepine therapy (usually clonazepam) for at least six months, and often for many years. Among these patients, 146 (86 percent) found that this therapy completely or substantially controlled their sleep problems, 8 percent had adverse effects (such as morning sedation) requiring medication changes, 2 percent had relapses of alcohol or chemical abuse requiring hospitalization, and 2 percent at times misused their medications.[7] Other less common side effects include morning "hangover," dizziness, problems with balance, memory changes, or confusion.

Zolpidem, zaleplon, and eszopiclone are imidazopyridine medications that interact with a benzodiazepine receptor in the brain. Their advantages are rapid absorption, lack of active metabolites, and low risk for side effects or abuse. Zolpidem has been extensively studied, with an excellent therapeutic profile during nightly use for up to six months, although in our experience it can be used safely on a nightly basis for well over a year in patients who are carefully followed. The usual dose is 5–10 mg at bedtime, with doses of 15–20 mg at times being used. Zaleplon is an ultra-short-acting agent that is effective in promptly restoring sleep in patients having problems with nocturnal awakenings. The usual dose is 5–10 mg. Eszopiclone has been shown to be effective with minimal side effects after 12 months of nightly therapy. The usual dose is 2 or 3 mg at bedtime.

A new and intriguing option for treating insomnia is the melatonin receptor agonist ramelteon, approved almost two years ago by the FDA. It comes in an 8 mg dose. The jury is still out about its efficacy compared to the other classes of hypnotic medication.

The following antidepressant agents at times can also be effective in controlling insomnia: doxepin or amitriptyline (10–50+ mg), trimipramine (25–100 mg), imipramine (10–75+ mg), and trazodone (25–100+ mg). Doxepin in particular has clinical research backing its use in insomnia, with the other agents having far less research backing. Also, these agents carry risks of untoward effects.

INSOMNIA, AMBIEN THERAPY, SLEEP-EATING, AND SLEEP-DRIVING

As an example of how the therapy of insomnia can trigger a parasomnia, consider this report: Over a two-year period, seven physicians at our hospital's psychiatry clinic and sleep clinic found 19 cases of sleep-related eating disorder (SRED), and 2 sleep-driving cases induced by Ambien. When the patients stopped taking Ambien, the SRED disappeared. Our center has now identified more than 30 cases with this uncommon reaction to Ambien use. The identified factors for increasing the risk of SRED were: 1) 10–20 mg dose range, 2) being female, 3) having multiple medical and psychiatric disorders and being on multiple medications (for depression, anxiety, pain, hypertension, asthma, etc.), and 4) having restless legs syndrome. Therefore, please be aware of these risk factors and be prepared to discuss them, or even to initiate a discussion with the physician who is considering Ambien as the therapy of your insomnia. The fact remains, however, that only a small percentage of patients treated with Ambien will develop SRED, and a much smaller percentage will become sleep-drivers. My chapter on SRED will discuss this matter further.

MELATONIN

Melatonin is a naturally occurring hormone secreted by the pineal gland, a small gland in the center of the brain. Its secretion is controlled by the light-dark cycle. Although often thought of as a "sleep" hormone, it is actually a "darkness" hormone, secreted at sunset in animals that are day-active (like humans) as well as in nocturnal species. Melatonin levels are ten times higher in our bodies at night than during the day.

This hormone doesn't put you to sleep—although it may get dark out at 7:00 p.m. and melatonin levels may be high, you probably won't be ready for bed for hours. However, there is a lot of interest in how melatonin can function as a sleep aid, and it has proven useful in certain specific conditions. For example, melatonin is effective in improving sleep in a subgroup of elderly patients with insomnia who have low melatonin levels. But the studies are inconsistent: Many studies have shown no difference in the time to sleep onset or the duration of sleep between those taking melatonin and those taking a placebo.

Melatonin may be an effective therapy in resetting the circadian clock among people who have circadian-rhythm disorders. I will discuss this further in chapter 6.

One of the tricky things about taking any sort of dietary supplement is that they're not regulated by the Food and Drug Administration (FDA).

Manufacturers call melatonin a dietary supplement, so they don't have to pass the same rigorous standards for safety and purity that prescription drugs do, and no safe dosage has been determined. Although few serious side effects have been associated with melatonin thus far, don't believe that the labels that say "all natural" mean "safe."

HERBAL REMEDIES

Americans have shown a growing interest in "alternative therapies," spending billions of dollars annually on botanical remedies. However, currently there is insufficient scientific knowledge on the efficacy and safety of herbal treatments of insomnia. Valerian and kava are probably the most commonly promoted herbal remedies of insomnia in America, with valerian being more rigorously evaluated. One study found it as effective as a benzodiazepine medication over six weeks of treatment. However, as with supplements, herbal remedies lack standardization and governmental regulation. There are also possible interactions between herbs and drugs, so I would urge anyone who is taking an herbal product for sleep to inform his or her physician about this, since herbal products are often left out of the medical-history-taking process. The same interaction issue applies to over-the-counter medications.

INSOMNIA AND THE PARASOMNIA CONNECTION

Insomnia can be related to just about any parasomnia. A person who suffers from restless legs syndrome has trouble falling (or staying) asleep because of the uncomfortable or even painful feelings in the legs. A person with sleep-related violence, sleepwalking, or night terrors may be so stressed about the condition that it makes it difficult to sleep, and the condition can greatly disturb the sleep of the bed partner or roommate, who may also become stressed and develop sleep problems. Night-eating syndrome (NES) is a condition similar to a parasomnia (for example, SRED), in which the person awakens and then compulsively and excessively eats food before returning to sleep. Sleep-maintenance insomnia and a parasomnia-like condition are thus linked. (NES will be discussed in the chapter on SRED.)

In these cases, the parasomnia is usually the primary diagnosis—if we can get that under control, then the secondary insomnia should go away, too.

The treatment for insomnia can also cause a parasomnia. As I mentioned, it's been shown that zolpidem (Ambien) treatment for insomnia can trigger sleepwalking and sleep-related eating disorder.

MOST EFFECTIVE TREATMENTS[8,9,10]

Prescription medications combined with behavioral treatments may provide the best results for controlling chronic insomnia. Medications are best for providing immediate relief, while behavioral treatments gain effectiveness over the course of several weeks or months. Cognitive-behavioral therapy, which has been shown to be highly effective in major research studies, is a prime option in managing chronic insomnia for those people who can gain access to this form of therapy. It requires specialized training, is generally only available in urban areas, is labor-intensive, and is costly if not covered by your health plan. This therapy has multiple components: stimulus control, sleep restriction, relaxation training, cognitive therapy, and sleep-hygiene education.

Stimulus control involves removing stimuli in the bedroom that could interfere with sleep, such as computers, television sets, video games, telephones, checkbooks, etc. The goal is to reassociate the bed and bedroom with sleep. Sleep restriction is meant to maximize the amount of sleep you get when you are in bed, since many people with insomnia spend too much time in bed, tossing and turning, and thereby they associate the bed with sleeplessness. Relaxation training can help "tone down" the nervous system and help prepare oneself for sleep. It involves techniques for reducing body tension and mental imagery that dampens intrusive thoughts while enhancing pleasant thoughts. Cognitive psychotherapy aims at identifying and modifying erroneous or maladaptive thoughts (and associated behaviors) about sleep, insomnia, and next-day consequences. Sleep-hygiene education is usually a commonsense list of do's and don'ts concerning diet, alchohol, caffeine, exercise, noise/light/temperature in the bedroom, etc. Sometimes a simple change such as turning the alarm clock backward so the giant digital numbers don't taunt you all night can help keep insomnia at bay.

Whatever you do, don't let insomnia become an accepted way of life. Seek medical attention if the problem becomes frequent; there may be underlying medical or psychological conditions causing your sleep problem, or you may just need some help getting back to a healthful sleep routine. There are many remedies available to remind you of what it feels like to get a good night's sleep.

Chapter 3

OBSTRUCTIVE SLEEP APNEA

*Suffocating Repeatedly
in Your Sleep*

When football legend Reggie White died in 2004 at the age of 43, the country was shocked. How could a healthy athlete die in his sleep? An autopsy revealed that he had sarcoidosis in the lungs and heart—a potentially serious but rarely fatal condition—and that his obstructive sleep apnea, diagnosed in 1990, may have been a major contributing factor in his death.

Obstructive sleep apnea (OSA) is by far the most common reason people are studied overnight in a sleep lab. It's a pretty scary-sounding problem: For it to qualify as OSA, a person must *stop breathing* (the definition of *apnea*) for at least ten seconds, a minimum of five times per hour throughout the night. Yet this is not a deadly problem by itself. Why?

When you fall asleep, the muscles around your upper airway relax, which narrows down the respiratory passages. That is fine for most people. Your upper airway will remain open enough to allow air to pass through to the lungs and then back out uneventfully. However, if you have OSA, your respiratory passages collapse and become too constricted during sleep, and airflow stops completely *(apnea)* or partially *(hypopnea)*. Most times with sleep apnea, your breathing will stop for 10 to 30 seconds, though it may go on for more than a minute. Your blood oxygen saturation will likely decline as you cease to breathe. But the problem is easily solved as the oxygen-deprived brain sounds an alarm—the arousal trigger—forcing you to wake up, jolting the upper airway to open up and allow proper airflow to resume; you stay alive and then immediately resume sleep, and soon the abnormal apnea process repeats itself. Of course, this "solution" has an obvious drawback: You're waking up briefly ("arousing") 50 or 100 times per night (or per hour!), and the repeated disruption of sleep has major consequences, such as chronic sleep deprivation and daytime sleepiness.

Most of the time, a person with OSA is not aware of these obstructive episodes at all. He or she may find out from the bed partner, who has witnessed the choking or gasping sounds or the lapses in breathing—or who most often complains of the loud snoring. Or the affected person may simply come in because of feeling tired or excessively sleepy during the daytime without knowing why. When they find out that it's because they've actually been arousing from sleep numerous times all through the night, it finally makes a lot of sense.

Typically, a person with OSA is also a loud snorer. This happens because the air passing through the narrow airway causes the soft tissues in the mouth and throat to vibrate from turbulent airflow. Early in sleep, the person will begin snoring, getting progressively louder, and the bed partner may then also become aware of the apneas or hypopneas because of the periodic lapses in snoring. Snorting sounds are not uncommon when the apnea ends, and before the snoring resumes.

The obstruction can occur anywhere in the upper respiratory path, but usually is within the back of the mouth or throat. A person may have more than one area of obstruction, including the nose. Apnea events happen more often in lighter NREM stages and especially in REM sleep, and not very frequently in the deeper stages of NREM sleep.

The severity of a person's OSA is measured by what's called an apnea-hypopnea index (AHI): the number of apneas and hypopneas per hour of sleep. This is also known as the respiratory disturbance index (RDI). Typically, an AHI/RDI score below 5 is not considered abnormal in adults. Determining the level of severity of OSA can be a tricky issue, since it involves not only the AHI/RDI score but also the extent of any associated decrease in oxygen levels in the bloodstream during sleep that is called "oxygen desaturation." In general, an AHI/RDI score above 30 suggests at least a moderately severe level of OSA. However, having a high RDI does not always lead to adverse daytime consequences. Some people with a high RDI have minimal daytime sleepiness, whereas other people with a modestly elevated RDI can have considerable daytime sleepiness.

How Do We Measure Sleep?

In a sleep lab, we use various tests and readings to analyze a person's sleep patterns and behaviors. The *polysomnogram,* or "sleep polygraph test," measures the most important physiological activities surrounding our sleep. Here are some of the more common assessments.

Electroencephalogram (EEG)

An electroencephalograph measures the brain's electrical activity. We attach several electrodes (metal discs) to a person's scalp, and the machine produces an electroencephalogram (EEG) that depicts the brain's ongoing electrical activity: a reading that shows a continuous line marked by peaks and valleys emerging with different frequencies that correspond to different parts of the sleep cycle. The test measures the frequency and the amplitude of the waves. Delta waves are the slowest at 1–3 cycles per second, followed by theta waves at 4–7 CPS, then alpha activity at 8–12 CPS, ramping up to the runs of sleep spindles and beta and gamma, fast and superfast activity. The classification of sleep into its component stages depends in large part on the predominant frequency of the EEG waves.

Electromyogram (EMG)

The electromyogram records electrical activity in skeletal muscles. Both muscle tone and muscle twitching can be detected. We expect to see muscles become inactive during REM sleep. During other stages, the EMG may detect teeth-grinding, leg and arm movements, and face twitches. This procedure is also done with electrodes on the skin, usually the chin *(submental)* muscles and the shin *(anterior tibialis)* muscles.

Electrooculogram (EOG)

The electrooculogram measures eye movement activity. It's useful in detecting the rapid eye movements of REM sleep and the excessive, high-voltage, slow and medium frequency activity found with a condition called *Prozac eyes.*

Electrocardiogram (ECG)

The electrocardiogram continuously records your heart's electrical activity through the sleep study.

Respiratory Monitoring

This has two components utilizing different recording devices: the monitoring of airflow (to detect sleep apnea) and the monitoring of "respiratory effort" in the chest wall and abdominal muscles (to distinguish between OSA, in which there is great effort expended to overcome the apnea, and central sleep apnea, in which there is a lack of effort, as the brain periodically shuts down the drive to breathe). A small device called a *pulse oximeter* is placed on a finger to measure ongoing oxygen content in the blood.

Continuous Positive Airway Pressure (CPAP) Titration

For patients with obstructive sleep apnea, this test in the sleep lab determines how much air pressure is needed to keep the person's airway open and unblocked. The patient wears a mask over the nose (or nose and mouth) connected to a tube attached to the CPAP machine that delivers the air pressure.

Audiovisual Monitoring

An infrared camera placed above the bed continuously monitors sleep behaviors and sounds, in a "time-synchronized" manner with the polysomnogram, in order to document any abnormalities and pinpoint the responsible sleep stage.

Multiple Sleep Latency Test (MSLT)

Patients taking a Multiple Sleep Latency Test have histories of excessive daytime sleepiness. They spend the preceding night monitored in the sleep lab to determine that they obtained at least five hours of sleep. After they awaken spontaneously in the morning, they get out of bed, and then return to the sleep-lab bed every two hours for 20 minutes with the bedroom door closed and the lights out. This test is meant to determine how long on average it takes the patient to fall asleep during each of the four or five nap opportunities, and also to examine the sleep stages present during those naps. (For example, the presence of REM sleep in at least two of the naps, together with demonstrated excessive sleepiness, can help document the diagnosis of narcolepsy.) This test helps to distinguish between *objective sleepiness* and *subjective sleepiness*—the latter of which may be physical tiredness or lack of energy but is not a true inability to stay awake.

SYMPTOMS OF OSA

- snoring
- gasping or choking for air during sleep
- labored breathing and snorting sounds during sleep
- repeated cessation of breathing during sleep
- nocturia (frequent need to urinate at night)
- unrefreshing sleep
- dry mouth on awakening
- daytime sleepiness
- depression or irritability
- impaired concentration and/or memory
- high blood pressure[1]

WEIGHT AND OBSTRUCTIVE SLEEP APNEA

In a population-based study of 6,132 participants 40 years old or older, researchers determined that the more severe OSA is, the more likely it is that the sufferer is obese.

- Among those with mild OSA, 16.4 percent had a normal body mass index (less than 25), 38.1 percent were overweight (BMI of 25 to less than 30), and 45.5 percent were obese (BMI of 30 or more).
- Among those with moderate OSA, 13.9 percent had a normal BMI, 32.6 percent were overweight, and 53.5 percent were obese.

• Among those with severe OSA, 10.2 percent had a normal BMI, 28.7 percent were overweight, and 61.1 percent were obese.

WHO GETS OSA?

About twice as many men as women have OSA, and the predominant trait most sufferers have is excess body weight. The risk of OSA increases as a person gets heavier—there is an extremely high rate of OSA among those who are considered morbidly obese. Excess fat on the sides of the airway makes the passage even narrower to start with, so it takes less to collapse it completely. A short, thick neck is a strong predictor of more obstructive episodes (AHIs) per hour. A neck circumference of 17.5 inches or more is highly correlated with the risk of OSA.

You don't have to be overweight to get OSA, however. Some people with normal or below-normal weight have other structural problems that make them prone to OSA. For example, those with large tongues or receding jawlines have an increased risk.

It occurs in people of all ages, but the risk increases during the transition from middle age to older age.

Other factors that may put a person at increased risk for OSA include:

• menopause
• smoking
• endocrine disorders (hypothyroidism, acromegaly)
• asthma
• epilepsy
• Down syndrome (in adulthood)
• enlarged tonsils and adenoids (the most important risk factor for OSA in children)
• history of a broken nose or deviated nasal septum

Factors that may make OSA worse include:

• alcohol consumption before sleep
• weight gain

- rhinitis (allergic or irritant-caused inflammation of nasal passages, sinuses, and eyes)

HOW COMMON IS IT?

Different studies have shown widely differing figures. The results vary partly because some count only symptomatic OSA, and some count asymptomatic OSA. (Asymptomatic OSA means that apneas are detectable on sleep tests but the person doesn't complain of any sleep-related problems.)

The prevalence has been reported as being anywhere from 4 percent of men and 2 percent of women to 24 percent of men and 9 percent of women. (The first set of figures is the one most often cited.) It's likely that OSA is on the rise in the United States because obesity is on the rise.

WHY IS IT DANGEROUS?

Because people with OSA have such fragmented sleep, they often experience daytime fatigue and sleepiness. In clinical tests, it's been shown that those with OSA perform poorly in comparison to control subjects on tests of vigilance and attention. "Not only is their ability to remain awake in monotonous situations impaired but their ability to maintain attention in more stimulating conditions is also affected," writes a team of researchers in the sleep and respiratory unit of a university hospital in France.[2]

Excessive sleepiness, as we've discussed, is at the root of many motor vehicle accidents and workplace accidents. OSA also puts people at higher risk for clinical depression, and has its own set of adverse consequences.

OSA—on account of its repeated nocturnal hypoxic (low oxygen) states, and perhaps also its high carbon dioxide states—is also probably linked with long-term cardiovascular consequences, such as cardiac arrhythmias, pulmonary hypertension, myocardial infarction, congestive heart failure, and stroke. However, the data gathered to date are not yet definitive, because of confounding variables and other factors, and important additional studies are currently in progress. It's been difficult to prove a cause-and-effect relationship between OSA and cardiovascular disease; we know that patients with heart disease have high rates of OSA, but it's unclear whether the OSA was part of the cause, and if so, how much of a contributing factor it was. Nevertheless, patients with hard-to-control congestive heart failure (CHF) should be carefully questioned and evaluated for OSA, since control of any associated OSA can facilitate the control of CHF.

Because most people with OSA are also overweight, it may be difficult to separate the possible causes; obesity alone clearly puts a person at higher risk for heart disease, and OSA by different mechanisms may increase that risk. The same is true of diabetes mellitus: While people with diabetes mellitus (types I and II) have higher rates of OSA, we don't know for certain whether the OSA is part of the cause or whether obesity is at the root of both conditions.

It has been established, however, that OSA can be part of the cause of systemic hypertension. At least half of all patients with OSA are hypertensive, and studies have shown that the correlation between OSA and hypertension exists independently of age, sex, and obesity.

A *metabolic syndrome* that features insulin resistance, diabetes mellitus (especially type II), weight gain, and various other problems is both a promoter of OSA and an increasingly recognized consequence of OSA. Therefore, OSA should clearly be recognized as a systemic illness rather than just a localized problem in the respiratory tract during sleep.

DANGEROUS DRIVING

Researchers in Virginia compared the driving records of patients with OSA to those without. They discovered that the automobile-accident rate of the patients with OSA was 2.6 times the accident rate of all licensed drivers in the state of Virginia, and that 24 percent of patients with OSA reported falling asleep at least once per week while driving.[3]

TESTING FOR OSA

Those suspected of having OSA will need to be evaluated by a sleep professional and then be monitored in a sleep lab overnight as already explained. None of this monitoring is invasive and shouldn't cause pain or discomfort.

The technician and physician are watching for specific things. First, they need to make sure that during episodes of apnea or hypopnea, you're actually making an effort to breathe. If you are making an effort, you have *obstructive sleep apnea*. If not, you have *central apnea,* a much less common form of sleep apnea where the brain is at fault—you stop breathing because your brain just isn't making any effort to breathe for periods of time. *Mixed apnea* is a combination of the two: You aren't making the effort to breathe some of the time, but when you do make the effort again, you have an airway obstruction that stops you from breathing. Throughout this process of observation, your blood oxygen levels are being closely watched.

TESTING AT HOME

There are now some do-it-yourself oximetry tests for OSA. These measure oxygen levels during sleep, but the jury is still out on how valid they are. Patients cannot purchase these tests on their own. Some sleep doctors will hook up the equipment and give instructions, then send the patient home to sleep. Many insurers, quite reasonably, will not cover treatment of OSA until the patient has undergone full sleep-lab testing. A major concern among sleep doctors is the following: Many people with sleep-disordered breathing will have multiple breathing-related arousals on a nightly basis—but negligible decreases in blood oxygen tension. Therefore, a home oximetry device that is the sole screening test for OSA could produce a "normal" result even though the person has disrupted sleep from repeated OSA-induced arousals and complaints of daytime fatigue and sleepiness. Even though treatment is warranted in such a scenario, the normal home oximetry screening wouldn't allow such treatment unless a formal sleep-lab study was performed.

TREATMENT FOR OSA

Let's take a look at the medical and surgical treatments for OSA. First, some lifestyle changes may help. Avoid alcohol and sedatives before going to bed since they can worsen the OSA. Changing your sleeping position may help, since OSA is generally worse in the supine position. Creative methods such as sewing a pocket to insert a tennis ball in the back of a nightshirt to keep you from sleeping on your back can help too.

If you are overweight and have OSA, losing weight can certainly help—but in general, it is very difficult, if not impossible, to achieve until your OSA is treated. If you are tired and sleepy from OSA, it is virtually impossible to muster the energy to exercise, eat properly, and do all the other things necessary to lose weight.

But in most cases, lifestyle changes alone won't have much impact on OSA, and so you will need some kind of medical intervention.

Medical (Nonsurgical) Options

There are useful devices to treat OSA that carry far less risk than surgery.

CONTINUOUS POSITIVE AIRWAY PRESSURE (CPAP)

This is the most commonly recommended and most effective nonsurgical treatment for OSA, and multiple studies have conclusively demonstrated the effectiveness of CPAP in controlling OSA. Established, reputable companies sell the machines and provide ongoing service for patients, and most health insurance companies cover the cost of a CPAP machine. The treatment includes a plastic mask that is fitted over the nose (or nose and mouth) and attached by hose to a machine that delivers pressurized air into the back of the throat throughout the night, keeping the airway open. Nasal-CPAP is the most common treatment, but there are two other variations: bi-level positive air pressure (BiPAP) and auto-adjusting (also called auto-titrating) positive air pressure (APAP, also called "smart-PAP"). The BiPAP delivers two levels of pressure: higher pressure when the person inhales, lower pressure when the person exhales. BiPAP is often used in patients who cannot tolerate or otherwise don't like CPAP and is also used in treating some patients during sleep who suffer from restrictive lung disease (i.e., pulmonary fibrosis, or disorders of structures surrounding the lungs, such as obesity and kyphoscoliosis), which can be associated with *hypercapnia* (increased levels of carbon dioxide). The APAP is meant to sense changing pressure needs throughout sleep.

To determine how much pressure is needed, patients must undergo a CPAP titration test. You may have this done on the same night as your PSG. This is called a *split-night study;* the first part of the night is devoted to the diagnostic PSG and the second part is devoted to the CPAP titration test. During this test, the technician fits the patient with a mask (or nasal cushions) and while looking at the sleep tracings is able to adjust the levels of air pressure until finding the right pressure to eliminate the apneas and any breathing-related arousals, which will then also restore normal levels of oxygen tension in the bloodstream.

The biggest problem with CPAP treatment is compliance. It appears that less than half of those who are advised to use CPAP actually do so every night, all night. There are many reasons for this, of course—it may feel uncomfortable, it may feel embarrassing if you have a bed partner, and some people say they feel like they're suffocating when wearing a mask over the nose. However, like many things, CPAP may just take some getting used to. An experienced sleep-center nurse can be an invaluable resource in helping you feel more comfortable, and there are CPAP support groups. (You can contact the National Sleep Foundation at www.sleepfoundation.org for further information.)

The original CPAP models, introduced in 1981, have been improved upon considerably. There are now humidified devices that blow warm, moist air, which can take away much of the discomfort if a patient experiences nasal stuffiness, nosebleeds, or dry or sore throat. There are also "nasal pillows" if the patient doesn't like wearing a mask. Also BiPAP is a promising alternative therapy in some patients.

It's been shown that those with OSA who are treated with CPAP or BiPAP have reduced daytime sleepiness, improved quality of life, less risk of motor vehicle accidents, and less depression. It has at least a 70 percent success rate in treating OSA when patients use the device properly, and receive proper medical follow-up, as officially described.[4, 5, 6, 7]

DEPRESSION DUE TO OSA

This case comes from my psychiatry clinic. A 38-year-old man came in with a one-year history of depression that he could attribute only to "not being able to complete many of my tasks at work and at home because of poor attention and sleepiness." He also reported being a snorer, as confirmed by his wife, who joined us during the second part of our interview. He had been seeing a psychotherapist, who could not identify any other basis for my patient's depression. I ordered an overnight sleep-lab study that documented OSA, with 32 apneic events per hour and lack of delta NREM sleep, with only mild decrease in oxygen tension, all of which responded fully to nasal CPAP therapy. When I saw him after one week, and then again three months later, he was very pleased to report that his depression had fully lifted and that he had regained full attentiveness during the day, was no longer sleepy, had increased energy, and could complete his tasks in a timely fashion. We agreed that antidepressant medication therapy was not needed, and he entered the termination phase of psychotherapy. This case shows how OSA not only can cause snoring, but it can also lead to psychological depression, poor attention, daytime sleepiness, and decreased functioning at work and at home—with full resolution after proper diagnosis and appropriate treatment with CPAP.

ORAL APPLIANCES

Another treatment that may work for (mild) OSA is the use of an oral appliance (OA). A person can wear a fitted plastic device like a sports mouth guard over the teeth during sleep, meant to keep the jaw forward, elevate the soft palate, or keep the tongue forward. The therapeutic result involves opening the upper airway. This may help reduce snoring, but its success rate is much lower than CPAP therapy in mild, moderate, or severe OSA patients. However, its compliance rate is better: Reported patients used the oral appliance 77 percent of nights at the one-year follow-up, and about 85 percent of patients will continue to use their oral appliances for at least a year. It is typically recommended only if the patient can't tolerate CPAP therapy and has mild OSA. Oral appliances can cause jaw and tooth discomfort and excess salivation, but these side effects are uncommon. They can also cause minor movement of the teeth in some patients after prolonged use.

Oral appliances should be fitted by qualified dental personnel who are trained and experienced in the overall care of oral health. Patients with OSA being treated with an OA should return for regular follow-up visits with the dental specialist to discuss how the treatment is working and to be sure the device is not deteriorating. If the patient's OSA symptoms become worse, they'll need to have another sleep study.[8, 9]

Surgical Options[10]

Surgery is a controversial area in the management of OSA, and an excessive number of unwarranted and therapeutically unproven surgical procedures are performed across the country (and world) every year. I would recommend that a patient obtain a second opinion before undergoing surgery for OSA, except when the purpose of the surgery is straightforward and obvious, such as enlarged tonsils, deviated nasal septum, or some other obvious structural deformity; or when you are working with an established university or university-affiliated sleep center that has an established protocol and, ideally, conducts clinical research. It's also important to know that you will have careful monitoring after any surgery to make sure the OSA has been resolved, and to check for side effects such as infection, bleeding, breathing problems, and swallowing problems.

Besides having a careful physical exam you should have a specific test that is highly recommended for examining the upper airway: fiberoptic nasopharyngoscopy, which allows for a direct examination of the entire upper airway from the nose to the larynx. Also, your physician should X-ray your facial-skeletal anatomy (called a *lateral cephalometric radiograph*). Doing a CT (computerized tomography) or MRI (magnetic resonance imaging) of your airway would also be ideal.

Your physician should explain the risks, both short-term and long-term, and the probability of those risks to you (and ideally also to your spouse). You should try to establish a good rapport not only with the surgeon but also with the nurse and sleep techs. Keep the communication lines open.

Let's look at the most common surgical procedures for OSA.

Uvulopalatopharyngoplasty (UPPP)

This has been the most common OSA surgery over the past 25 years. With UPPP, excess tissue is removed from the back of the throat and the soft palate (the uvula) to widen the airway passage. The tonsils are also removed. This procedure may be done with a scalpel, or with laser treatment, in which case it's known as *laser-assisted uvulopalatoplasty* (LAUP). However, the American Academy of Sleep Medicine does not support the use of LAUP in treating OSA because of insufficient evidence of success. A uvulopalatal flap procedure is a modification of the UPPP procedure. In clinical studies of UPPP, up to 37 percent of those who initially showed improvement relapsed in two to eight years. The true rate of short-term and long-term benefits in controlling OSA is still not known. However, we should keep in mind that this procedure will reduce or eliminate snoring, which is one of the hallmarks of OSA, and so it may mask one of the prominent symptoms of OSA without improving it. There are various complications to this procedure, including pain and difficulty swallowing.

Temperature-controlled Radio-frequency (RF) Tongue Base Reduction

Through use of a needle electrode, radio-frequency energy is delivered to the upper airway tissue to shrink it and stiffen it, opening up the airway, resulting in improvement of OSA. Current studies have shown that, although somewhat effective, this procedure is best viewed as an adjunctive treatment in combination with other surgical interventions.

Jaw and Facial Surgery

This type of surgery may help those patients with OSA who have small or receding jaws. In the procedure, the upper and lower jaws are cut and repositioned to pull them forward. This allows more room for the tongue, so it's less likely for the tongue to block the airway. It's not a simple surgery, nor is it inexpensive, and the patient's jaw will be wired shut for about a month afterward, requiring a liquid diet. However, it does have a high success rate. Very informally, we call these procedures "facial moving-around surgery." These are serious, invasive procedures that do hold promise in treating some patients with OSA. The tricky question is, Who will benefit? Sometimes the answer is clear, in patients with prominent anatomical defects, but other times there are subtle abnormalities that may not be linked with the sleep-disordered breathing. I cannot overemphasize that you should be carefully evaluated for this type of procedure by an experienced surgeon knowledgeable about OSA, one who has performed these procedures many times and can explain the risks, and ideally, one who practices at an academic institution and accredited sleep-disorders center that conducts outcome studies.

Nasal Surgery

Nasal surgery is sometimes used as a part of overall OSA treatment, but it's rarely effective alone. It may be done to repair a deviated septum *(septoplasty)* or remove blockages in the nose, which may help other treatments be more effective.

Tracheostomy

Originally, doctors didn't have many other options but to treat OSA with a tracheostomy (which in "the old days" was often lifesaving). In this procedure, a breathing tube is inserted directly into the *trachea* (or "windpipe") through an opening in the front of the neck below the larynx, thus bypassing the upper airway obstruction. The opening is closed with a valve during the day and opened before the person goes to sleep. That works, but it's a truly extreme measure. There are many risks and complications involved, not the least of which is social stigma and concerns about appearance. The health risks include increased risk for lung infections, buildup of scar tissue, bleeding, and speech difficulty. This was a recommended treatment for OSA until 1980. Nowadays, it's used as a temporary measure, or only when OSA is severe and no other treatment has worked.

CLOSE CALLS WITH OSA

A morbidly obese 55-year-old man couldn't tolerate his CPAP machine, so he was switched to a BiPAP machine that "adequately" controlled his OSA. Unfortunately, over the next four years, he gained more than 220 pounds and never had his OSA checked again.[11] His treatment needed adjustment, but he didn't know that. His family described a man who was falling apart—he was increasingly sleepy during the day and was paranoid, anxious, and moody. This came to a frightening head one night when he lay down to sleep and reached to put on his BiPAP mask and somehow didn't realize he'd also grabbed the gun he kept on his nightstand. When he tried to fasten the mask on his head, the gun fired and the man accidentally shot himself in the scalp—luckily, it wasn't a lethal wound and he was able to dial 911.

The doctors studying him determined that his undertreated OSA was causing long-term fragmented sleep and repeated nocturnal hypoxia. This, in turn, severely hurt his quality of life. His memory and vigilance were impaired, and his psychological state was tenuous. After his gunshot wound was treated, he went into respiratory failure (most likely due to pneumonia), and doctors decided to perform a tracheostomy—which ended his OSA. They report that his mood and all cognitive symptoms improved once his OSA was treated.

This story emphasizes the very serious complications of OSA, and serves as a reminder that patients should always attend follow-up appointments to monitor their conditions. OSA can worsen over time, particularly if a person gains weight, and treatments may need to be modified. In this man's case, it's very possible that he was aware of his daytime symptoms but didn't attribute them to his OSA because he thought that as long as he followed his instructions with the BiPAP machine, he was treating it properly. Of course, it also serves as a reminder of why deadly weapons should never be kept out in the open in a bedroom!

In a frightening, similar story, a 54-year-old woman with untreated severe OSA had been experiencing parasomnias, including sleep-driving, for five years before her daughter persuaded her to seek help.[12] What finally brought her into a sleep lab? She chopped up her cat on a cutting board while she was asleep one night. Since being treated with CPAP therapy for her OSA, her parasomnia symptoms have disappeared.

SEEKING HELP FOR OSA

If you suspect that you or someone you know has OSA, it's important to see a sleep professional. You may need to visit your primary care doctor for a referral, depending on your health insurance. Take this seriously, because OSA can literally put your life at risk and there are effective solutions for dealing with it.

Try to choose a sleep professional who devotes at least half of his or her clinical practice to the field of sleep medicine, and thereby demonstrates a focus and commitment to sleep and its disorders. Too many physicians who claim to be sleep doctors in reality spend less than 10 percent of their practice time in a sleep clinic, and they often sign off on the scoring and limited interpretation of the polysomnogram by a technician without a careful review of the sleep record, and without obtaining a full clinical sleep-wake history from the patient (and spouse) in order to correlate the sleep-lab findings with the clinical sleep complaints. As my colleague Mark Mahowald has been saying for years, "Sleep is not just a pulmonary function." By that he means that even if a sleep problem involves OSA, there are often far-reaching consequences that affect and detract from a person's quality of life. Therefore, treatment of the patient with OSA involves far more than ordering a CPAP machine. Other health issues must be considered, along with the psychological well-being of the patient.

OSA is too often overlooked as an annoyance because of the snoring rather than viewed as the very serious problem it is, both on its own and as a "gateway" to other sleep disorders. If you or a loved one have the symptoms described in this chapter, get tested and get it under control.

Chapter 4

RESTLESS LEGS SYNDROME

The Agony of Jittery Drowsiness

R estless legs syndrome (RLS) is the third-most–commonly reported sleep disorder, and one of the most common causes of severe insomnia. Including its mild forms, this irritating disorder affects between 10 and 15 percent of the population. Although it's classified separately as a movement disorder, it's important to note that all sleep-related movement disorders are technically parasomnias. Together they comprise the movement and behavioral disorders related to sleep.

The way people describe their symptoms is often imprecise and can vary considerably, but the common denominator is an unpleasant sensation in the legs. Sufferers may say they feel a "creepy, crawly" sensation in their legs when they're trying to sleep or when they're just resting quietly. Or they describe extreme discomfort, pain, pulling, searing, boring, or deep itching sensations in the legs. Some of my patients have described "electric ants" or bugs running up and down the legs. These distressing symptoms are typically relieved only by movement or stimulation of the legs. Doctors call the symptoms *paresthesias* (abnormal sensations) or *dysesthesias* (unpleasant abnormal sensations). For some people the feelings are a minor annoyance, and for others they're sheer torture. In all cases, there's an urge to move the legs, and getting the legs moving eases the symptoms temporarily.

It's different from leg cramping or discomfort that doesn't include an impulse to move. It's also different from hypnic jerks, which are involuntary spasms at sleep onset that don't involve the prior desire to move. Many conditions that cause pain can be mistaken for RLS, but if you really have RLS, the primary sensation will be an insistent urge to move the legs that begins or worsens during periods of rest or inactivity.

The symptoms don't necessarily appear only when the person is trying to sleep—the feelings can occur anytime the person is at rest, such as when he or she is sitting to watch television or traveling in a car or plane. Generally, symptoms are worse at night and especially while lying down and feeling drowsy (what a cruel paradox, being sleepy while being restless!). It may affect only one side of the body, or both, and even the arms sometimes, and it tends to get worse with age.

The most frustrating part of this for most people is that the more they try to relax and ignore the symptoms, the worse it gets. Most choose to get up and walk around for relief, though sometimes just stretching and massaging the legs while still lying down is effective. People try all sorts of creative techniques for relief; some take hot baths, some use wraps or compresses, some stomp their feet on the ground. Any of these actions can bring at least some relief, but the symptoms typically return within a minute or two of lying down again.

Shelly's Story

I was about 31 when I was put on Elavil (amitriptyline) for depression. It came with a nasty side effect, though: As I was drifting off to sleep, I would shake my legs so hard it looked like I was chasing rabbits! My legs felt jittery—you know how it feels when you feel fidgety, so you wiggle a foot or cross your leg over the other and shake it up and down? Turn the intensity up to max, just when your mind and body are exhausted, and that's what I had.

I put up with it for four or five months because my depressive symptoms were so severe. As bad as my leg issues were, it still didn't compare to the pain going on inside my head at the time, so I waited it out as long as I could. Meanwhile, my patient husband lost sleep because I kicked him all throughout the night. He told me that I spooned up against him and "jackhammered" his back with both my knees—rapidly and repeatedly, often throughout the night.

In April, I finally went in for a sleep study. I was diagnosed with restless legs syndrome and periodic limb movements during sleep. They told me my sleep was interrupted almost 50 times that night by my legs—vigorous enough to kick off the electrodes, and the technician had to replace them twice. I had no idea how often I was awakened through the night—I wasn't alert enough to remember by the morning—but my brain waves during the study clued the sleep specialists in; I was repeatedly arousing myself from sleep involuntarily throughout the night. No wonder I was always so tired!

Going off the Elavil helped alleviate most of my symptoms. I still get jittery legs sometimes, but it's not as severe. My current remedy is to get my husband to massage my low back and sometimes my legs. It relieves a lot of the crawly, got-to-shake-my-legs sensation. Sometimes visualization helps, too. I imagine I'm traveling right within the muscles of my legs and giving them a massage at the cellular level and increasing the blood flow, encouraging the muscles to relax. It may sound silly, but it seems to work for me.

WHO GETS RLS?[1]

Although it can strike people of any age, RLS is most commonly reported among people over the age of 40, and is classified as *early onset* if it strikes before the person is 45 years old. It's 1.5 to 2 times more common in women than in men.

Many people who are later diagnosed with RLS report having symptoms in childhood, and there's mounting evidence to suggest that it's often misdiagnosed among children as attention deficit/hyperactivity disorder or "growing pains."[2, 3, 4, 5]

In more than half of cases, primary RLS runs in families. If you have an immediate family member with RLS, your risk is three to six times greater of developing the disorder. Current exciting research shows that there may be a gene marker for RLS that explains this familial connection.

WHAT CAUSES RLS?

RLS may be *primary* or *secondary*. Primary RLS means that there is no underlying disorder causing it, whereas secondary RLS means that another condition (medical or neurological) or a medication is really at the root of it. In the latter case, treating the underlying condition or stopping an offending medication should resolve the RLS. Although the majority of cases of RLS are primary (with presumed problems with the brain's iron and dopamine systems), there are some known causes and associations with other conditions:

- Iron deficiency and anemia are known to increase the risk of RLS.
- Several types of medications may bring on or aggravate RLS, including many cold and allergy medications, antinausea medications, antidepressants (except Wellbutrin [buproprion], a dopamine-enhancing medication that may actually help), antiseizure drugs, and antipsychotic drugs.
- About 20–40 percent of those with chronic kidney failure on dialysis have RLS. The RLS symptoms disappear after a successful kidney transplant. The strong link between kidney disease and RLS is still greatly under-recognized, especially among kidney specialists.
- Peripheral neuropathy, Parkinson's disease, Lyme-disease myelitis, and diabetes mellitus have all been linked with RLS.
- Caffeine, tobacco, and nicotine can aggravate or trigger RLS symptoms.

There is no evidence that RLS is linked with any sort of underlying psychological problems, nor does it signal the onset of any other neurological disorder. However, because the distressing symptoms and resulting sleep deprivation of RLS can be chronic, it can induce depressive and anxiety disorders.

Pregnant Women and RLS

About 26 percent of pregnant women get RLS, and it gets worse in the second and third trimesters. Ten percent of them already had the disorder before they got pregnant, but it generally gets much worse during pregnancy. The remaining 16 percent didn't have the disorder before, and some worry that it's going to harm the baby, or that it's a sign of something wrong. It isn't. RLS in the mother does not pose any known risk for the baby.

The problem that requires treatment is when the mother is losing significant amounts of sleep most nights because of the RLS. Dopamine medications, which are the most-recommended treatments for RLS and the best option, are not considered safe during pregnancy. But clonazepam carries fewer risks. As with all disorders during pregnancy, the risk of the disorder itself has to be weighed against the risk of the treatment. Thus far, there are 75 cases in the sleep literature about pregnant women with RLS taking clonazepam. One of those women had a baby with a birth defect. That's about the same risk as in the general population, which suggests that the medication doesn't have known harmful effects on a baby. To feel safer about making recommendations, however, doctors like to wait until there are about 500 cases officially reported, to make sure the sample size is big enough to detect complications.

One of the strange things about RLS in pregnancy, though, is that it usually goes away (or at least lessens) during the last couple of weeks, or days, before the woman goes into labor. This is one of the reasons it's not clear what's causing the RLS: Many scientists believe it's low iron levels, but those last couple of weeks are the time when pregnant women have the least iron. The babies have used up most of the mother's available iron. So, it's still possible that iron plays a role, but it's unclear in what way. It is known that iron is a necessary cofactor in the synthesis of dopamine, and dopamine deficiency during sleepiness is the apparent basis for RLS symptoms in those predisposed. So relatively low iron could contribute to low dopamine that promotes RLS symptoms during drowsiness.

The blood tests for finding out if iron deficiency could be aggravating RLS symptoms are blood (serum) iron level and ferritin level tests. Ferritin is the blood transporter protein for iron, and it has a low level when there are low iron levels. (There is no need to have a higher level of transporter protein when there is no more iron to transport.)

WHAT'S THE BIG DEAL?

On first glance, it may sound like RLS is just an annoyance, but for those with moderate or severe cases, RLS can be maddening, exquisitely painful, and seriously disrupting sleep quality and duration. In the early stages of RLS, it is unusual for a person to have episodes every night, but eventually those who develop moderate to severe cases can have symptoms virtually every night. It may be very difficult for the person to fall asleep and/or stay asleep. Therefore, it can leave the person feeling fatigued or even totally wiped out the next day—which, of course, leads to all the potential problems I've mentioned in earlier chapters, such as depression, impaired concentration, and impaired coordination.

PERIODIC LIMB MOVEMENTS OF SLEEP

Between 80 and 90 percent of those who are diagnosed with RLS also have a finding recorded in the sleep lab called *periodic limb movements of sleep* (PLMs), which may or may not be associated with symptoms of sleep disruption. (This is currently a controversial area in the field of sleep medicine and research, since the presence of PLMs—even large amounts of them every night—does not per se indicate a disorder called *periodic limb movement disorder.*

PLMs describe the involuntary movements of the limbs during sleep. Typically, the movement occurs in the feet, ankles, and legs, but movements can also occur in the arms and hands. Episodes typically last less than 5 seconds at a time and occur every 15 to 40 seconds. The episodes must involve four distinct movements—such as a twitch of the big toe followed by an ankle movement, then jerks of the knee and hip. It's not the same as a normal body shift or simple stretch during sleep. Most times, people with PLMs, or those few who have bona fide PLMD, aren't aware of their movements during sleep.

PITY THE PARTNER

One of the problems with RLS and PLMs is that they can both significantly disrupt the bed partner's sleep. In fact, while those with PLMs often have no real repercussions from it, the partner is more often affected. The movements can make it very difficult for a partner to fall asleep or stay asleep. In turn, they will develop an environmental sleep disorder.

SYNDROMES CONFUSED
WITH RLS AND PLMS

The "painful legs and moving toes" syndrome that sometimes follows spinal cord injury can be confused with RLS. Nocturnal leg cramp disorder may also be mistaken for RLS. Also, akathisia is a side effect of antipsychotic medications in which the patient generally feels restless (even when standing) and moves around excessively, both during the day and at night. In contrast, RLS becomes worse in the evening, when the patient is seated or lying down, and is relieved by standing up and walking.

Sleep starts may be confused with PLMs. Sleep starts—also known as hypnic jerks—generally occur only during the transition from wakefulness to sleep, and are shorter, less complex movements. They last 20–100 milliseconds and have no repetitive pattern. Sleep starts are very common and are considered normal. Another condition, *myoclonic epilepsy,* involves repetitive leg jerks or arm jerks during sleep, with some of the movements being quite pronounced.

HOW RLS IS DIAGNOSED

RLS is diagnosed clinically during the interview with a physician. In most cases, a person with RLS doesn't need to be studied in a sleep lab. Although there are formal diagnostic criteria contained in the ICSD-2, which I will describe, a recent collaborative study from Italy and the United States found that a single screening question for RLS had a 95 percent sensitivity in diagnosing RLS (meaning that almost everyone with RLS would answer yes to this question).[6] This screening question is: "When you try to relax in the evening or sleep at night, do you ever have unpleasant, restless feelings in your legs that can be relieved by walking or movement?" People who answered no had a 98 percent probability of not having RLS. Those who answered yes had a 62 percent probability of having true RLS, meaning that there were some "false positive" responses that required follow-up questions and use of these formal diagnostic criteria that define RLS:

- There is an urge to move the legs, usually accompanied or caused by uncomfortable and unpleasant sensations in the legs.
- The urge to move or the unpleasant sensations begin or worsen during periods of rest or inactivity such as lying or sitting.
- The urge to move or the unpleasant sensations are partially or totally relieved by movement, such as walking or stretching, at least as long as the activity continues.
- The urge to move or the unpleasant sensations are worse, or occur only, in the evening or night.

Ask yourself the screening question for RLS if you are having trouble falling asleep at night, and if the answer is yes, then consult your doctor, who should be able to question you further. Your doctor may want to perform tests to rule out underlying conditions such as iron deficiency, anemia, and kidney dysfunction or to check for nerve and muscle damage. Your physician may also wish to refer you to a sleep specialist for management of your RLS, depending on his or her level of knowledge about RLS.

COGNITIVE IMPAIRMENT AND RLS[7]

A troubling issue is how to diagnose and treat people with dementia and other forms of cognitive impairment (such as from stroke) who have RLS but who cannot communicate that they have the symptoms. I observed this in my last year of residency as a psychiatrist, when I spent half a day a week at a nursing home, and many people who stayed there had to be strapped into their wheelchairs or otherwise confined because there was a risk that they would wander or fall and hurt themselves, or because they had violent or aggressive tendencies due to Alzheimer's or another disease. Most of them were noncommunicative, at least to the extent that they couldn't accurately describe what they were feeling.

If they wailed, screamed, or moaned much of the time, we would try to determine if they were in significant physical pain. If so, they could be given painkillers. If not, we would generally assume that the negative vocalizations, or aggression, were part of the dementia or due to cognitive impairment following a stroke or other medical event. Then the person might be given a "major tranquilizer"—that is, antipsychotic medication.

What's troublesome for me to realize now is that antipsychotic medications tend to make RLS worse, and that being confined to a chair or bed is torment for a person with RLS. In retrospect, I am sure that some of these people certainly did have RLS, so unknowingly, doctors probably aggravated it while thinking they were doing something to help.

This applies not only to people in nursing homes but to anyone who has cognitive impairment and an inability to communicate symptoms. Considering our aging population, it's something that people should be more aware of when dealing with their elderly relatives. Furthermore, the elderly are at greater risk for iron deficiency and kidney disease, which are risk factors for triggering RLS.

There are two main questions to determine whether RLS is likely in a given scenario: Do the agitation and vocalizations get worse when the person is at rest (sitting for long periods or lying down)? And does it get worse in the evening? If the answer to both of these questions is yes, it's a good idea for doctors to try the "dopamine challenge," which is to try treating the person with a dopamine-enhancing medication to see if the agitation lessens. If it does, there's a strong chance that the person has RLS.

The medical literature to date on this topic remains unfortunately sparse. Much more work needs to be done in this area.

TREATMENTS FOR RLS

This is the good news: Relief is at hand, since most people with RLS can be substantially or fully controlled with medication therapy. However, sometimes RLS can be very difficult to control, in which case multiple medications at rather high doses are required—assuming that iron deficiency is ruled out. The crucial first step is accurate diagnosis, followed by establishing a good working relationship and open communication between the doctor and the patient (and spouse/family).

There are four main classes of medications used to treat RLS.

1. Dopamine-enhancing medications comprise the foundation of treatment, since several lines of evidence point to dopamine deficiency

in RLS. (I should point out that there are several chemical sub-types of the neurotransmitter dopamine—it is a complicated story, just like the serotonin story is very complicated—and so there can be different categories of dopamine-deficient disorders; for example, RLS is *not* linked to Parkinson's disease.) Pramipexole (Mirapex), and ropinirole (Requip) are the most commonly prescribed and effective medications for RLS, with L-dopa/carbidopa (Sinemet) falling into disfavor because of greater side effects. Ropinirole, however, is the only FDA-approved drug for treating moderately severe to severe RLS. As just indicated, people with RLS are not more prone than other people to develop Parkinson's disease. Occasionally, these dopamine-enhancing medications may relieve symptoms at night but cause symptoms to reappear the following day (a side effect called *augmentation*); in that case, the same medication may also need to be taken during the daytime, particularly in anticipation of periods of inactivity, such as taking a long car ride or going to see a movie.

2. Benzodiazepines such as clonazepam or temazepam. These medications are muscle relaxants working at the spinal-cord level that also have sleep-promoting and anti-anxiety effects mediated in the brain. They may not fully blunt your leg sensations or block the periodic leg jerking, but they can help you sleep despite them, by increasing the "arousal threshold"—that is, they make it harder to be disturbed by any disruptive stimulus.

3. Opiates such as propoxyphene (Darvon), codeine, oxycodone, tramadol, and even methadone in severe cases. These narcotic medications control the pain that can be a disabling feature of RLS. Long-term therapy is effective and safe, although sleep apnea may emerge in some patients.[8]

4. Anticonvulsants such as carbamazepine or gabapentin. Anticonvulsants are used to prevent or manage seizures, and may also help reduce restlessness associated with RLS. They are also sometimes used with patients who report feeling pain with their RLS.

Most patients will find relief by using one of these medications, though some experimentation may be necessary to find the right treatment and the right dosage. Sometimes a patient responds best to a combination of two or more medications, particularly a dopaminergic agent combined with an opiate.

It's understandable that some patients (and doctors) are concerned about the abuse and addiction risks of benzodiazepenes and opiates, but the actual incidence of abuse, tolerance, and addiction appears to be quite low among those with RLS. If the RLS is severe, the patient should maintain long-term treatment with close, ongoing clinical follow-up to assure that proper control of RLS is established and to monitor the dosages and make sure that misuse or abuse of medication does not occur. Also, the medications used to treat RLS can cause dry mouth and related complications.[9] See your dentist if these problems occur. These medications can cause various other complications, such as nausea and dizziness, so have your doctor inform you about them before starting therapy.

NIGHTTIME EXERCISE

While it's generally advised that people should exercise early in the day to promote better sleep, some people with RLS report that moderate exercise at night helps to decrease their symptoms.

SEEKING HELP FOR RLS

It's important to have a doctor review your medication list and check for underlying conditions (such as iron deficiency) that may be causing RLS. If there are no underlying problems found, depending on the severity of the symptoms, long-term medication therapy may be appropriate. RLS rarely disappears or improves over time. Although it's not dangerous by itself, it can seriously disrupt sleep quality and lead to excessive daytime sleepiness. Lifestyle changes such as cutting out caffeine and reducing alcohol use can make a significant difference, whether in conjunction with medication or not. Getting a bigger bed to avoid disrupting the bed partner probably wouldn't hurt, either! There are other "tricks of the trade" in controlling RLS.[10]

It's important to watch for the signs of sleep-related eating disorder (SRED) as well, because RLS does put people (particularly women) at higher risk for this disorder. See chapter 12 for a full explanation of SRED.

Although there is greater awareness of RLS now, it seems that doctors are not always quick to diagnose or treat it. In a recent study in Austria, researchers determined during face-to-face interviews with more than 700 people that 10 percent of the general population between the ages of 50 and 89 had RLS (14.2 percent in women, 6.6 percent in men), but none of them were being treated for it—not one, even though about 22 percent of them had severe cases, and two-thirds had moderate-to-severe cases.[11]

So don't assume that just because your regular doctor doesn't diagnose you, it means you don't have it. If your problems with restless legs are bothering you and it's disturbing your sleep, make sure to talk to a specialist and find some relief.

•

Chapter 5

HYPERSOMNIAS

Excessive Sleepiness,
Daytime Dysfunction

Hypersomnia is a disorder in which sufferers have frequent, excessive daytime sleepiness that isn't attributed to a lack of sleep or interrupted sleep at night, or when they consistently sleep longer than normal at night.

A person with excessive daytime sleepiness will feel the strong urge to sleep at inappropriate times—like at work, at school, driving a car, or in the middle of a conversation or an activity—despite getting a normal night's sleep. One of my patients put his problem this way: "It is hard work to stay awake."

The most common cause of hypersomnia is a rebound effect from intentional sleep deprivation due to the work-related and social commitments most people have these days. We get 20 percent less sleep than previous generations, and there is no evidence that our needs are less. Nonintentional sleep deprivation is almost always due to an underlying sleep disorder, most commonly obstructive sleep apnea or narcolepsy.

As with insomnia, but to a lesser extent, depression is a major cause of hypersomnia. People who are clinically depressed may sleep too little, or at times too much, which makes sleep a fascinating barometer of a person's mood.

LONG SLEEPERS

A long sleeper is someone who consistently sleeps more than most other people in his or her age group. Although the sleep is long, it's basically normal in every other way. To qualify as a long sleeper, an adult must show a consistent pattern of 10 to 12 hours of sleep per night for at least a week.

A person should not be diagnosed as a long sleeper if there is any medical or mental explanation for the extra sleep—for example, some people experience symptoms of hypersomnia following an illness such as infectious mononucleosis or viral pneumonia; even after the illness has passed, patients may still feel the aftereffects of an increased need for sleep that can last for weeks or months. They may also mistake fatigue for sleepiness and take excessive naps to try to make up for it. This is not true primary hypersomnia, though. Likewise, if a person is sleeping longer hours due to medication, or to another sleep disorder such as sleep apnea or narcolepsy, or to depression (particularly bipolar depression and "winter depression"), then the person is not intrinsically a long sleeper.

According to surveys, about 2 percent of men and 1.5 percent of women sleep at least 10 hours per night, but they may not all be long sleepers—some certainly have underlying conditions causing them to sleep more than the norm.

In most cases, as long as these long sleepers get the sleep they require, they wake up easily and function fine—but if they try to limit their sleep to more "normal" durations, they experience excessive daytime sleepiness.

This biological trait begins in childhood. By itself, it's a bit controversial whether or not it's a "disorder," as there are no abnormalities in the quality of sleep or related health concerns. Doctors may prescribe stimulants in an attempt to limit a patient's sleep duration, but this is often not effective. In effect, the sleep itself is its own treatment: If these people allow themselves to sleep extra hours on a regular basis, it will "cure" them of excessive daytime sleepiness. Consequently, they must adjust their lives accordingly and play with the sleep cards that Nature dealt to them.

The problem, however, is usually of a social or work-related nature. Long sleepers may be embarrassed that others perceive them as lazy, or they may feel that they can't get enough done during the day.

SHORT SLEEPERS

Short sleepers are at the other end of the spectrum, generally requiring five hours of sleep or less per 24 hours for adults, or for children, a minimum of three hours less than is age appropriate. They tend to get a normal amount of the deepest sleep—stages 3 and 4 delta sleep—but a reduced amount of stage 2 and REM sleep. They show few interruptions of sleep, so they're very effective sleepers! Short sleepers don't report any impairment in daytime functioning or mood, and they don't purposely restrict their sleep—they just naturally awaken after a relatively short period of sleep and function well during the daytime.

This is different from insomnia—either primary insomnia or insomnia from other conditions or medications—which result in negative daytime consequences. Short sleep is a long-term biological trait that begins in childhood or young adulthood and often runs in families.

A short sleeper at times may worry that it's not normal to sleep so little, and occasionally may even worry about dying prematurely from "wearing out" his or her system from getting so little sleep. (One patient worried that "my heart will shut down when I'm 60 because of all the extra heartbeats from being awake so much all my life.") However, as long as the person shows no adverse effects from having short total sleep time, current evidence indicates that treatment is not needed and there is no cause for concern about adverse health consequences or shortened life span.

YOU SLEEP LIKE A FLY

Fruit flies sleep a lot like humans do, which is why flies are useful insects to study in sleep research. Sleep-deprived fruit flies show impaired performance, and they compensate with deeper and longer sleep the next day, just like people do.

It's been a curiosity among scientists why some people are short sleepers. How is it that a small percentage of the population can get by just fine on four hours of sleep, while most of us need twice that much to function well? Scientists from the University of Wisconsin at Madison led by Giulio Tononi, M.D., Ph.D., decided to figure it out, using a species of fruit fly, the *Drosophila melanogaster,* as their model. This type of fly is used widely in genetic studies because it is abundant, its whole genome has been mapped, and mutants of all types are readily available for research.

After a four-year exhaustive study of 9,000 flies, they discovered a mutant line that they called *minisleep* flies: flies that sleep for one-third the normal amount of time yet are not impaired by their sleep deprivation. These minisleep flies all had the same amino acid mutation of a particular gene (the Shaker gene). This gene is responsible for controlling the flow of potassium into cells. Human studies are now needed to determine if we have a comparable gene responsible for sleep regulation.

NARCOLEPSY

Narcolepsy is a condition in which the barriers between sleep and wakefulness become repeatedly blurred.

Imagine this: You're sitting in a class, jotting down lecture notes. A few minutes later, you open your eyes—*Hey, I didn't mean to fall asleep!*—and you notice that you've written several lines of scribbles. You look around, hoping no one else noticed that you nodded off, and try to force yourself to keep your eyes open.

You're sleepy most of the time, even though you sleep at least seven or eight hours a night. You sometimes have a strong urge to fall asleep while standing up, or while shopping, or at a stoplight while driving. It's an irresistible urge much of the time, and you often have to pull over to the side of the road, or find a bench to nap on. People tell you that you're lazy, that you should just pay attention and quit "letting yourself" sleep all the time, that you should drink coffee and exercise more and snap out of it. You're not sure whether to believe them. No one needs this much sleep . . . *Maybe I* am *just lazy.*

You go out to the diner with a group of friends, and someone tells a joke. You laugh—and lose muscle control. Your head drops down, your eyelids droop, and you slump down in your seat with your jaw hanging open. Friends think you're kidding around, but you can't control it. You're alert, but you're unable to move or talk. About a minute later, you regain muscle tone and brush it off, wishing they'd find something else to joke about.

That night, within minutes of lying down, you begin hallucinating. You hear voices that seem to be in the room, yet muffled. It feels intensely real, and even though you're not sure exactly what's going on, it's terrifying. The next morning when you wake up, your body is paralyzed. You're completely alert, but you can't move or call out, and you feel there's an evil presence in the room holding you down. This may be brief, or it may last for several minutes. *Am I going to suffocate? Am I dying?*

The list of possible misdiagnoses is staggering. Sufferers and doctors alike may believe it's a stroke, multiple sclerosis, panic attacks, schizophrenia, a thyroid condition . . . Narcolepsy is uncommon enough that much of the general medical community doesn't understand it, yet common enough that they should. Several studies have shown that it often takes many years for a patient to be diagnosed correctly with narcolepsy. One UK survey of 219 narcoleptics revealed that the average time between the onset of symptoms and a correct diagnosis was 10.5 years, though the time to proper diagnosis has decreased in recent times.[1]

One Family's Story

In 1997, our sleep center worked with a family who had a history of narcolepsy. It was a novel case in which both biological parents were afflicted, along with all three of their children. What's fascinating in the context of this book is that the first member of the family who came to our attention in the late 1980s was the 23-year-old daughter—who was referred on account of sleep-related eating disorder, a parasomnia that we were just beginning to understand at that time.

During my initial interview with this daughter, it was apparent that she had a major problem with daytime sleepiness, besides her nightly problem with sleep-eating. We eventually documented narcolepsy in all five family members, an unprecedented finding.

The mother also had a long-standing sleep-related eating disorder, and one night she had a dream that she was cooking and preparing for a dinner party later on that night. When she woke up, she saw the dining room table set and the food ready to be served, and she was dressed for the occasion—and soon became vexed when none of the guests had arrived, since it was already 6:30 and the people had been told to come at 6. But then she finally woke up enough to realize that it was 6:30 a.m., not p.m., and there was no dinner party planned after all—"I knew I had done it again."

At any rate, the mother described the following scene that she witnessed from the porch of the family farmhouse one day: It was late morning on a sunny day, and she was enjoying herself on the porch as she watched her husband and 19-year-old son go about their work driving separate tractors. Suddenly, she noticed that both of them had experienced an almost simultaneous sleep attack, and as they were slumped over the wheels, their tractors started zigzagging across the farm field.

At first she was amused by this unusual scene, but then she became concerned that the two tractors would crash into each other. Then she became very worried as the two tractors headed to the highway by their farm, which they eventually crossed—and shortly thereafter, a large truck came speeding down the highway as the two tractors were stuck in a ditch on the other side of the road. The blast of the passing truck woke up a bewildered narcoleptic father and son, and then they started trying to put the pieces together of what had just happened. As it turned out, that episode helped convince the rest of the family to come to our sleep center to be evaluated soon after their daughter had come to us.

Characteristics of Narcolepsy

There are four distinctive markers of narcolepsy:

1. **Excessive daytime sleepiness.** Sufferers have trouble staying awake, even in active environments. During a structured sleep-laboratory test named the *multiple sleep latency test* (MSLT), they fall asleep quickly when given opportunities to nap. This is not related to insufficient sleep at night (they must get at least six hours in order for the next-day MSLT to be valid) or attributable to another medical or psychological condition. Generally, narcoleptics fall asleep several times a day, for periods of just a few minutes to an hour or two. After a nap, the drowsiness subsides for an hour or more, but can come back anytime after that. Most people with narcolepsy can feel a buildup to a "sleep attack" coming on, even if it's quick.

They may also experience *microsleeps*—moments of sleep when the person is not aware they've slept.

2. **Cataplexy.** Cataplexy is the sudden loss of muscle tone during wakefulness, triggered by a strong emotion. A cataplectic attack may be brought on by laughter, fright, anger, shock, or sadness; or running, sexual activity, dancing, or similar catalysts. Although not all people with narcolepsy have cataplexy, when it is present, it's a very telling and specific marker for the diagnosis. Mild cataplexy may include a feeling of weakness in the knees, the sagging of facial muscles, the head dropping forward, and the arms dropping to the sides. Speech may also become slurred. Severe cataplexy can affect seemingly all voluntary muscles, causing the person to drop to the floor without warning, likely injuring him- or herself or those nearby. This can easily be mistaken for a seizure, but the main difference is that afterward, those with cataplexy can express a complete memory and awareness of their surroundings—they were alert the whole time, just unable to move. Most episodes last fewer than five minutes, but some can last more than an hour and result in sleep attacks, in which the person falls asleep while experiencing cataplexy and generally is fine upon awakening. This symptom is often the most embarrassing for narcoleptics, who may go to great lengths to avoid emotional situations or hide their faces in triggering situations (sitting in the back of the room, facing away from people). About 70 percent of narcoleptics experience cataplexy—and cataplexy is rarely seen in anyone who does not have narcolepsy.

3. **Hypnagogic and hypnopompic hallucinations.** Those with narcolepsy have unusual sleep architecture: They tend to move very rapidly into REM sleep, within minutes of falling asleep, and often right away begin dreaming. These "sleep onset" dreams can be very difficult to distinguish from reality. On the other hand, these dreamlike experiences may be vague and involve only one sense: sound or sight. They may be scary, or just odd but not disturbing. When they occur during the transition from wakefulness to sleep, they're called *hypnagogic hallucinations;* when they occur during the transition from sleep to wakefulness, they're called *hypnopompic hallucinations.* These events can involve misperceiving the bedside

surroundings, and are not limited to those with narcolepsy—they can occur in people who are sleep-deprived, or they may be linked with medications, alcohol, or sleep-schedule shifts.

4. **Sleep paralysis.** Sleep paralysis can also occur in the transition either from wakefulness to sleep or from sleep to wakefulness. It's a period of time when the brain is "awake" but the body isn't. Sleep paralysis may be accompanied by hallucinations, as well, or just a vague sense of doom and panic. Often, sufferers describe feeling as if an evil presence is sitting on their chests, suffocating them. The paralysis may be total, or it may involve just the arms, just the legs, or just certain other body parts. Often, patients describe trying to call out for help but being unable to make a noise. Sometimes they describe "out-of-body" sensations or a feeling of levitating. These events typically last a few minutes, but can be very brief.

Only 10 to 25 percent of narcoleptics experience all four of the classic symptoms, but most experience the first two. When excessive daytime sleepiness and cataplexy are both clearly present, a diagnosis of narcolepsy may be fairly easy to make. When cataplexy is absent, a doctor must confirm the diagnosis with overnight testing followed by an MSLT. On average, the patient should take no more than eight minutes to fall asleep during four or five nap opportunities (each separated by two hours), and must be found to enter REM sleep rapidly (within 20 minutes of sleep onset) during at least two of these naps.

Narcoleptics often have very disturbed sleep throughout the night, enforcing the idea that narcolepsy is no good at respecting boundaries. Repeated awakenings disrupt sleep and then sleep repeatedly intrudes into wakefulness. In addition, many narcoleptics demonstrate automatic behaviors that may be simple and brief or complex and prolonged, for which they subsequently have no memory.

Narcolepsy is highly associated with REM sleep dysfunction. Sleep-onset REM sleep attacks with vivid dreaming are common. Cataplexy and sleep paralysis reflect how the protective muscle paralysis in REM sleep that keeps us from acting out our dreams can inappropriately intrude into our waking lives. It appears in these cases that REM sleep and its parts are trying to invade wakefulness and show up on the scene as soon as possible once the person has fallen asleep, or persist when the person is awakening, or take over during strong emotional moments.

Narcolepsy is a major risk factor for another type of REM sleep motor dysfunction—REM sleep behavior disorder (RBD).[2] This isn't surprising, because both disorders involve an abnormal loss of boundaries between wakefulness and REM sleep, and both disorders also involve abnormal control of muscle tone, muscle twitching, and behaviors. In fact, narcolepsy and RBD are bizarre and twisted mirror images of each other: Whereas a narcoleptic patient while being awake will suddenly develop brief, reversible attacks of paralysis (i.e., cataplexy) when he should have muscle tone, a patient with RBD will *not* be paralyzed in REM sleep as he should be, but instead will have muscle tone, muscle twitching, and inappropriate behaviors while dreaming. So a person with narcolepsy and RBD has a severe double whammy, day and night. Recent studies from the U.K. and Italy have identified a high prevalence of RBD in narcoleptic patients, ranging from 36 percent to 52 percent, but with the rate of RBD ranging from 68 percent to 79 percent in those narcoleptics who also suffered from cataplexy.[3, 4] Furthermore, a potential cause or aggravator for RBD in narcolepsy concerns the medications used to treat cataplexy, namely tricyclic antidepressants, SSRIs (such as fluoxetine [Prozac]), and venlafaxine (Effexor).

Until narcolepsy is under control, sufferers shouldn't drive. The National Institute of Neurological Disorders and Stroke estimates that people with untreated narcoleptic symptoms are involved in ten times as many automobile crashes as the rest of the population.[5] I've heard convincing evidence of this among my patients, too—narcoleptics often report having multiple car accidents before seeking treatment. (Often, the threat of losing their licenses is what brings them into the sleep lab.) Many say that there have been times when they were unaware of how they got home or when they changed lanes, or that they nodded off despite turning on the radio or having someone in the car talking to them. These dangerous automatic behaviors obviously pose a major danger to the person with narcolepsy and to anyone else who happens to be on the road.

Although not caused by a psychiatric disorder, those with narcolepsy often suffer serious psychological consequences. They may be teased by peers for their sleepiness, napping, or "strange" cataplexies; they may suffer from poor self-esteem because of poor academic or work performance; they may become very shy and withdrawn because they feel like outsiders; they may develop high anxiety and depression; they may have trouble forming social bonds because they feel judged by others . . . In short, narcolepsy is a disorder with serious potential physical, psychological, and social ramifications.

It can also lead to other sleep disorders. Narcolepsy is associated with obesity, which increases the risk for obstructive sleep apnea. This, in turn, can make the daytime sleepiness even worse, leading to more frequent sleep attacks!

How Prevalent Is the Problem?

The prevalence of narcolepsy in the United States is approximately .05 percent, or 1 in 2,000 people, similar to the prevalence of Parkinson's disease and multiple sclerosis. There is an increased familial risk, but children of narcoleptics still have only a 1–2 percent chance of developing the disorder. Narcolepsy may begin at any time but usually starts during puberty or early adulthood. It affects men and women about equally.

How Is It Treated?[6]

Treating narcolepsy and cataplexy may require more than one medication, and several options are available.

- **Modafinil (Provigil).** Approved for treatment of narcoleptic excessive daytime sleepiness by the FDA in 1999, this is the first non-amphetamine, non-methylphenidate stimulant medication to be approved to help narcoleptics stay awake. It carries a lower risk for side effects and abuse than amphetamines, and up to five refills per prescription are allowed (versus 0 refills for amphetamines)— therefore, it is a more user-friendly medication. Modafinil is currently the initial treatment of choice for narcolepsy. The side effects are generally minimal, and long-term use doesn't appear to require progressively higher doses to maintain control of excessive sleepiness.
- **Stimulants.** Traditional central-nervous-system stimulant medications, such as methylphenidate (Ritalin), dextroamphetamine, and

amphetamine salts, have long been prescribed to help narcoleptics control sleepiness and maintain alertness. A doctor must carefully monitor their use, though, since side effects can be troublesome or even serious, particularly at higher doses; they include appetite suppression, mood and mental changes, and increased blood pressure. Patients may develop a tolerance to these medications. The drug may stop working, and increasing doses over time may not continue to be effective, although it generally is. Sometimes a drug from this class is combined with modafinil for maximum benefit.

- **Antidepressants.** To control cataplexy, three types of antidepressants may be used on account of their chemical profile: selective serotonin reuptake inhibitors (SSRIs), particularly fluoxetine (Prozac); venlafaxine (Effexor); and tricyclic antidepressants (clomipramine, imipramine). All three have been proven to decrease cataplexy episodes, generally with mild side effects and low risk of tolerance. Note: The use of antidepressants to treat cataplexy doesn't mean that the person has clinical depression; it just means that these medications have chemical properties that work for both conditions. (A new medication called Xyrem—sodium oxybate—has been approved by the FDA for treating cataplexy, but it is very expensive, and generally not needed, since the other medications are quite effective and are much cheaper. Nevertheless, ongoing research is being conducted on Xyrem's possible additional benefits for narcolepsy.)

Recent Developments in Narcolepsy

One of the most exciting discoveries in sleep research in recent years concerns how the deficiency of a brain neurotransmitter that was discovered in 1998 named *hypocretin* (or *orexin*) can cause narcolepsy with cataplexy. Neurons containing hypocretin are located only in a small region of the hypothalamus, which is a major brain center for regulating our biological rhythms, including sleep and wakefulness and all the other basic body functions.

In 1999, researchers at Stanford University, led by Dr. Emmanuel Mignot, discovered that a mutation in the hypocretin receptor 2 gene caused narcolepsy in dogs.[7] Scientists quickly set out to find whether something similar was happening in humans. Since receptor dysfunction can be confirmed only at autopsy, the focus of research in humans with narcolepsy was on analyzing the levels of hypocretin in the cerebrospinal fluid (valid blood tests have not been developed yet). What they found was that 90 percent of those who have narcolepsy with cataplexy have an absence or very low level of hypocretin in their cerebrospinal fluid.

Besides all the fascinating scientific issues raised by this discovery, what might this discovery mean for those people affected by narcolepsy? It points the way for the clinical need to synthesize hypocretin and then be able to effectively and safely deliver it to narcoleptics so as to replace the lost amounts of this neurotransmitter in their brain. This should be the most effective and specific treatment for the sleepiness problem in narcolepsy, though development and release are years away. Perhaps cataplexy will be controlled as well.

It's not yet known why the hypocretin levels of narcoleptics are so low. It has been a long-standing suspicion in the medical community that narcolepsy is somehow tied to a malfunction in the immune system—it may actually be an autoimmune disease—but this has not yet been proven. Converging lines of evidence are lending stronger support to this theory. It may well be that in a manner similar to how diabetes mellitus and multiple sclerosis appear to get triggered, a viral illness or some other infection (even one that may seem to be trivial) or unknown immune-system trigger may induce an immune response that not only attacks the invading virus but also—by chemical "cross-reactivity"—attacks the hypocretin neurons, destroying them in their only location in the brain, which then wreaks havoc on the ability of the person to stay awake during the day while maintaining normal muscle tone.

We should think "smart bomb" (i.e., specific immune attack) in this scenario—a recent seminal autopsy study revealed that under the microscope, only the hypocretin neurons get knocked out (with scar tissue replacing the lost cells) and not the melanin-containing neurons that surround the hypocretin neurons in the same region of the hypothalamus. In contrast, some people develop a secondary form of narcolepsy from a stroke or tumor affecting the lateral hypothalamus, which is not a "smart bomb" scenario, because all the neurons in that region of the brain get damaged, producing a mixed neurologic picture, with narcolepsy being one of several prominent clinical features.

Researchers have also discovered particular variants in a region of chromosome 6 known as the *HLA (human leukocyte antigens) complex* that are strongly linked with narcolepsy. More than 90 percent of Caucasian and Asian narcoleptics have these variants, in contrast to a 30 percent prevalence in the general population. This is the highest disease-related HLA prevalence across the board in all of medicine, strongly suggesting an autoimmune basis for this narcolepsy.

IDIOPATHIC HYPERSOMNIA[8]

When someone reports excessive daytime sleepiness but doesn't have cataplexy or other markers of narcolepsy, is not sleep-deprived by choice, and doesn't have a medical disorder or take medication that could cause sleepiness, then they may have *idiopathic hypersomnia. Idiopathic* means that it has no known cause. As with narcolepsy, this condition still must be documented objectively in the sleep lab by means of an overnight sleep study and a next-day multiple sleep latency test.

People who are afflicted with idiopathic hypersomnia can be put into one of two recognized categories: those who are long sleepers at night (i.e., more than ten hours) but still have excessive sleepiness while awake, and those who sleep between six and ten hours at night and have excessive daytime sleepiness. In both cases, a Multiple Sleep Latency Test must show that it takes the person less than eight minutes on average to fall asleep during nap opportunities, and there should be no more than one REM sleep period. The person must have excessive daytime sleepiness for at least three months. Stimulant therapy is the treatment of choice but may not be as effective as in the treatment of narcolepsy. As with any other sleep disorder, the practice of proper sleep hygiene—and a lifestyle dedicated to promoting good sleep—is crucial in facilitating a successful treatment outcome.

Also, some patients with obstructive sleep apnea whose nocturnal sleep has been fully restored by CPAP treatment still experience problematic daytime sleepiness, and they should be taking daytime stimulant medication. Provigil has been approved by the FDA for this use. Simply controlling the apnea at night may not be enough; a separate treatment of daytime sleepiness may be in order. Level of functioning and quality of life are critical priorities in the assessment of a treatment plan.

KLEINE-LEVIN SYNDROME[9]

In contrast to narcolepsy and idiopathic hypersomnia, which are generally daily hypersomnia disorders, the Kleine-Levin syndrome (KLS) is a prominent example of a recurrent or periodic hypersomnia disorder. KLS typically affects adolescent males, but females and all age groups can be affected. Episodes can last days to weeks and appear from once a year to ten times a year. In between these hypersomnia episodes, there is normal functioning in all spheres of the person's life. During an episode, sleep may dominate up to 18 hours of a person's daily life, so it can have a profoundly negative effect on one's ability to work, attend school, or maintain adequate relationships. Furthermore, besides excessive sleepiness, there are often other peculiar features such as hypersexuality (or deviant sexuality), binge-eating, irritable mood, aggressiveness, manic-like symptoms, and sometimes even mental confusion, hallucinations, and feelings of unreality.

Unfortunately, reliable treatments have not yet been found, although lithium carbonate—a prime treatment for bipolar (i.e., manic-depressive) disorder—may have the best chance of helping. A problem with assessing the benefit of treatment is that the episodes of KLS usually last days to weeks, so perhaps a so-called beneficial treatment really only reflects the natural course of KLS.

Chapter 6

CIRCADIAN-RHYTHM DISORDERS

Timing Is *Everything*

If you've ever traveled across time zones, you probably know what it feels like to have an out-of-whack circadian rhythm—better known as *jet lag*. You feel sleepy, out of sorts, a little confused, listless, and just "off." It's a temporary state, though, and is usually resolved within a week. However, some people feel jet-lagged all the time—without leaving their time zone.

They have a serious problem with their internal body clock (regulating sleep and wakefulness) being mismatched with the occupational and social demands of their life, which manifests as a circadian-rhythm disorder. (The word *circadian* refers to events related to the daily rotation of the Earth.) The mismatch can result either from external challenges to our circadian system, such as jet travel and shift work; from vulnerabilities in our internal clock (that runs too long or too short); or from lack of environmental time cues, as in blind people.

It is estimated that 25 percent of people who come to a sleep clinic with an insomnia complaint (especially trouble falling asleep) turn out to have a circadian-rhythm disorder (CRD) as the basis for their insomnia complaint. Therefore, CRD can masquerade as primary insomnia, just as restless legs syndrome can masquerade as primary insomnia.

We don't know the exact process by which the body maintains its circadian rhythm, but we know that the suprachiasmatic nucleus in the hypothalamus of the brain is the main timekeeper for regulating our instinctual behaviors such as sleep. Recent research has found that our natural circadian rhythm (without external cues) is 24.2 hours—very close, but not identical to, the 24-hour daily sun cycle. The suprachiasmatic nucleus is affected, or *entrained*, by certain external stimuli, known as *zeitgebers* (German for "time givers"). These zeitgebers send out the cues to keep our circadian clocks running on schedule.

The most potent cue, which can precisely synchronize our internal rhythms to a 24-hour day, is light, especially solar light: Our bodies depend on the rise and fall of the sun to get the signal of when it's daytime (time to wake up) and when it's night (time to sleep). The effect of light exposure on our circadian rhythms is critically dependent on its timing. Morning light shifts the sleep-wake clock earlier, whereas evening light shifts the clock later.

The second most potent cue for our circadian clock is the hormone melatonin, which is produced in the pineal gland in the center of the brain. Melatonin is known as the hormone of darkness, in that its level rises with the appearance of night. Therefore, light and melatonin have polar-opposite effects on our circadian rhythm, so manipulating them both often works well for people with CRDs. Other much less potent zeitgebers include meal times and social activities that can also be adjusted as part of an overall therapeutic plan.

Sleep isn't the only process that's marked by a circadian rhythm. Another basic function is body temperature, which rises and falls based on the commands of the internal clock. Your temperature should start declining at night, reaching its lowest point around 3 to 5 a.m. Then it begins to rise again around dawn. During the day it fluctuates from about 97 to 99 degrees, reaching its highest point in the evening (about 7 to 9 p.m.). The body temperature rhythm runs independently of sleep pattern, although they are usually synchronized. One of the indications of a circadian-rhythm disorder is when the sleep pattern and the body-temperature pattern are not synchronized.

There are two primary standard methods available for identifying a CRD, besides a clinical interview with a sleep specialist. First, the person who is having symptoms should keep daily sleep diaries for at least one week in order to track the timing and pattern of sleep and wakefulness. These diaries should include at least one weekend, and preferably two weekends, to check for any sleep differences between weekdays and weekends. The second method is *actigraphy,* in which the person wears an actigraph, a wristwatch-type device that records movement continuously throughout the day and night for one to three weeks, to figure out when the person is not moving (probably asleep) and when he or she is moving around (probably awake).

Two other forms of circadian-rhythm monitoring can be considered: First, the monitoring of *core body temperature* throughout a 24-hour period, which usually entails a rectal thermometer. Second is the 24-hour *melatonin phase typing* that is currently being used in research protocols, and its widespread clinical use may be virtually around the corner. The way it works is that people provide saliva samples into a simple capture device at regular intervals in order to assess the pattern of melatonin secretion throughout a 24-hour period. Therefore, reliable and valid tools are available for assessing our sleep-wake rhythms, identifying any CRD, and pointing the way to rational and effective treatments.

Finally, rating scales (such as found on the Horne-Ostberg Scale) are available for assessing our intrinsic morning/evening preferences. As you may know, the world can be divided into the "larks" (extreme morning types), "owls" (extreme evening or night types), and those in between. These scales can help shed light on a clinical complaint suggestive of a CRD.

DELAYED SLEEP PHASE DISORDER (DSPD)

The most common type of chronic circadian-rhythm disorder is delayed sleep phase disorder (DSPD). This is when a person's normal sleep-wake times are at least two hours later than conventional or socially acceptable times—usually falling asleep at 2 a.m. or later. The overall amount of sleep is normal, as is the quality of sleep, but the timing is off. Falling asleep earlier is difficult or impossible, and if it happens, the person will still wake up late (10 a.m. or later).

About 40 percent of people with this type of circadian-rhythm disorder have a family history of the disorder. It often begins in adolescence, and the average age of onset is 20. Among adolescents and young adults, the prevalence of this disorder is estimated at about 7–16 percent. Prevalence in the general population is unknown.

The problem isn't so much the disorder itself—the main issue is that the schedule and demands of modern society are not compatible with the sleep-wake timing habits of the affected person, so most people with DSPD don't get enough sleep. They may not be able to fall asleep until 3 a.m., but they have to get up for work at 8, so they're constantly building up a sleep debt and experiencing daytime sleepiness.

Keeping up good grades at school or performing well at work may be difficult. In turn, they may use caffeine and other stimulants during the day to counteract this sleepiness, but the activating effects can carry through into the evening hours, increasing the risk of delaying sleep onset even further. It can also lead to misuse or abuse of caffeine and other stimulants or alcohol and drugs. About half of those with DSPD also have clinical depression. However, we should also recognize that people with natural delayed sleep phase, or "owl" tendencies, can be involved in jobs that are a perfect fit with their circadian rhythms, such as musicians, freelance writers, night janitors, and so on, and so they do not develop a problem because they can work at night when they feel alert, and still get enough sleep because they don't have to wake up early with the rest of the world.

DSPD can be confused with insomnia, but it isn't the same thing. The person can indeed fall asleep and stay asleep—just not at a "normal" time.

People with DSPD are often subjected to a double challenge: Not only does their "phase delay" circadian rhythm keep them up too late at night, but also many people who stay up too late expose themselves to bright light while they are up late—and the bright light further delays their sleep onset.

Therefore, treating DSPD can use these methods, which can work together for the best shot at helping.

1. Bright light exposure early in the day (which helps reverse the phase delay of sleep onset during the subsequent night): Keep in mind that a cloudy day still contains enough sunlight (10,000 lux)

to help shift our clocks; people often mistakenly assume that a cloudy day has a negligible amount of sunlight, which is not true. People should go outdoors for 30 to 60 minutes as early as possible in the day. If it is raining or cold, exposure to a light box indoors can be utilized.

2. Dim light exposure in the evening for at least two hours before desired sleep time: This will "deactivate" the person and will enhance the dim-light melatonin onset that occurs with darkness in preparation for sleep.

3. Melatonin therapy: This can take two forms. First, recent research published in the journal *Sleep*[1] indicates that a low dose (even 0.3 mg) of melatonin taken as a pill between 5 and 6.5 hours before desired sleep onset works best for shifting the clock backward; second, taking a 3-mg dose of melatonin shortly before going to bed can have a hypnotic effect insofar as melatonin is the "dark hormone" that promotes sleep. The reason not to take a 3-mg dose five to seven hours before going to bed is that this may induce sleep too early in the evening.

Jennifer's Story

The neighbors notice the light on in my office when they go to sleep and when they wake up. "Do you ever sleep?" one finally asks. Well, yes, but she'd never know it. I sleep while she's at work.

I've always been a night person. As soon as I had any choice in the matter, in college, I made sure to avoid taking any morning classes. I didn't mind working and studying late into the evening, though; that's when I felt most alert.

Starting in my early twenties, my sleep habits went from odd to ludicrous. I couldn't fall asleep earlier than 5 a.m., and I usually stayed up until 8 or 9. Obviously this meant a day job was out of the question. That was okay with me because I wanted to work from home anyway, but work wasn't the only challenge. I could never make it to the bank or post office before they closed; I slept until 4 p.m. on most days and needed that first hour just to work on opening my eyes and feeling human. Much of the time, I woke up in a haze. My parents (and later, my husband) would tell me about conversations we had during my wake-transition stage, and I'd have no recollection of the conversations at all. (*I promised to tape a show for you? When? Oh, you know I wasn't really awake yet!*)

Being out of sync with the world grew tiresome, and I really missed seeing sunlight, which I was sure couldn't be healthy. So I tried to change my sleep cycle . . . again and again. Often I stayed up for two days straight, just so I'd be utterly exhausted at a "normal" time. My doctor prescribed sleeping pills, I bought a light box to use in the mornings, my husband tried waking me with blaring music or taking off my blankets in the winter . . . each attempt would work for a few days, but I could never keep it up.

I've learned to work around my natural rhythms, for the most part. I schedule appointments in the evening whenever possible, I grocery-shop at 24-hour stores in the middle of the night, and I set two alarms and enlist help from my family when I must do something during the day. I've improved my routine so that it's more "out-landish" than "ludicrous," but I have learned to accept that I'm just not programmed to function in a morning-bird world. Now if only someone would open a bank for delayed sleepers!

ADVANCED SLEEP PHASE DISORDER (ASPD)

The counterpart to delayed sleep phase disorder is advanced sleep phase disorder (ASPD). As you might imagine, this disorder is diagnosed when people have a strong propensity to go to sleep and awaken earlier than conventional times. They feel sleepy in the late afternoon or early evening, and wake spontaneously in the early morning and can't fall back asleep.

The prevalence of this disorder in the general population is unknown, but the risk increases with age. It's estimated to affect about 1 percent of middle-aged and older adults. Like DSPD, it's usually a long-term condition and can lead to substance-abuse problems if people use stimulant agents to treat sleepiness or sedative agents to treat early awakenings.

The treatment of ASPD is the exact opposite of that for DSPD: bright-light exposure in the evening (to activate the person and delay melatonin release) and melatonin administration in the morning (to delay the premature release of melatonin in the late afternoon or early evening).

A CEO's Story

During the late 1980s, a 60-year-old CEO of a major corporation came in to see me with his wife of the past 30 years. They had one complaint: Throughout his life, he would fall asleep too early in the evening, which posed a distinct problem whenever he hosted a corporate party in their home, a common occurrence as part of his prominent executive position.

He and his wife felt embarrassed that he retired early during his own parties because of irresistible sleepiness that overcame him during his usual bedtime hour. In fact, he quickly excused himself to the bedroom to prevent the greater embarrassment of falling asleep in a chair or on a couch in full view of his guests. Despite his desire to stay awake and continue to host his party, he would go to his bedroom between 8 and 9 p.m. and fall asleep at a time when his party was getting into full swing. He became concerned that somehow his image would be tarnished by these repeated premature departures from his own parties.

In contrast to a farmer with an advanced sleep phase circadian rhythm, whose intrinsic sleep biorhythm would be perfectly suited to his work, this CEO suffered from a clinical sleep disorder only in a particular social context—hosting a party. Otherwise, he was just a "lark" who went to sleep early and woke up early, a pattern without problem. Attending someone else's party was not usually an obvious problem, since he could leave early without excessive fanfare after being one of the first guests to arrive. Therefore, it is the mismatch between intrinsic circadian rhythms on the one hand and lifestyle and employment demands on the other hand that results in a clinical disorder. Timing is everything: The problem is in the mismatch of biorhythms and social demands.

IRREGULAR SLEEP-WAKE RHYTHM

Then there are those with chaotic sleep patterns, showing no apparent circadian rhythm at all. In one 24-hour period, they may sleep three or more times, varying times from day to day, though the total sleep time may be normal. Body temperature, too, may fluctuate with no clear pattern.

This probably results from an abnormality of the circadian clock that can be aggravated by external factors such as light exposure. A jeweler with irregular sleep-wake rhythm became the subject of a case study, published in *Chronobiology International,* because doctors realized that his work environment was the root of his problem.[2] The professional diamond-grading equipment he used forced him to work with a daylight intensity lamp, and he worked late hours—which meant that he was exposed to bright light at night. This cue gradually threw off his circadian clock, until he had to scale back to a part-time job due to excessive daytime sleepiness.

Doctors treated him with melatonin at night, plus bright-light therapy in the morning, and advised him to avoid bright light at night. Within one week, the man's sleep schedule stabilized and he reported significant improvement in his daytime sleepiness.

There may be a connection between irregular sleep-wake patterns and neurological disorders, as well. Patients with Alzheimer's dementia have high risks of circadian-rhythm disorders; one study in Spain found that 43 percent of patients with dementia had a sleep disorder, and of them, 67 percent had irregular sleep-wake patterns. A second study found that the (oral) temperature circadian rhythm was disturbed irregularly in 59 percent of patients with dementia, compared to 12.5 percent of the same age patients in the control group.[3]

SLEEP IN SPACE

NASA is concerned about the documented irregular circadian rhythms of astronauts, noting that significant sleep loss and disordered sleep schedules have been shown on space missions. "Such disruptions could have serious consequences on the effectiveness, health, and safety of astronaut crews, thus reducing the safety margin and increasing the chances of an accident or incident," reads one research paper.[4]

CIRCADIAN-RHYTHM SLEEP DISORDER, FREE-RUNNING TYPE

Free-running, or nonentrained, circadian-rhythm disorder means that the internal clock is not programmed to a normal cycle of about 24 hours but is much longer or much shorter instead. It is most common in people who are totally blind, and is thought to affect more than half of all totally blind people. In the absence of visual time cues (such as daylight), the body's free-running circadian rhythm tends to run on a longer period, more like 25 hours. Usually, this means progressively delayed sleep times for those with the disorder—though rarer, it also can run on a shorter cycle instead.

There have been few cases of sighted people with this type of disorder, but not much is known about why it occurs. There may be increased risk of this disorder among those with dementia and mental retardation.

In one case, a woman had temporary free-running circadian-rhythm disorder for 19 days following a car accident.[5] She had an aneurysm close to the suprachiasmatic nucleus, which seemed to temporarily destroy her internal clock. Instead of being on a 24-hour daily cycle, hers was more like 22.5 hours, so she went to sleep earlier each day. After the aneurysm was surgically repaired, she gradually went back to a normal sleep schedule.

JET-LAG DISORDER

This is a commonly experienced problem in our modern age of frequent and far-reaching travel that has promoted the concept of the world as a "global village." After rapid travel across time zones, our circadian rhythms can become out of phase with the local time. We can then experience insomnia, daytime fatigue and sleepiness, gastrointestinal upset or other physical symptoms. These gradually resolve as our internal clock catches up to the local time zone and a "circadian harmony" is restored.

The severity of symptoms depends on the number of time zones crossed and on the direction of travel. Travel eastward (which advances our clock) is usually more difficult than travel westward (which delays our clock) because for most people it is easier to delay falling asleep than to fall asleep earlier than usual. Also, people sometimes make their jet lag worse by using caffeine on an airplane when they should be trying to sleep, or they use alcohol as a sleep aid that will then backfire by giving the person a short, inefficient sleep.

SHIFT-WORK DISORDER

Working an overnight shift isn't just inconvenient—it's often dangerous. Workers who are required to work during normal sleeping hours are at high risk for *shift-work disorder,* which may result in insomnia, excessive sleepiness, impaired work performance, and safety hazards. This category includes those who work late-night shifts, early-morning shifts, and rotating shifts.

One study of 2,570 workers found that 10 percent of those working rotating or night schedules met criteria for shift-work sleep disorder.[6] The study showed that health concerns may be aggravated by shift work. Workers identified as having shift-work disorder had higher rates of ulcers, sleepiness-related accidents, absenteeism, depression, and missed family and social activities. It's also known that those with shift work have higher rates of cardiovascular disease and gastrointestinal disorders.

People often don't get used to shift work, either. Sleep-disorder symptoms don't ease over time; in fact, there's evidence that the longer a person works during normal sleep times, the worse the sleep disorder gets.

TREATMENTS[7]

Besides the treatments already mentioned for DSPD and ASPD, the following can be considered:

- **Chronotherapy.** This therapy is meant to capitalize on the person's intrinsic rhythm, to "embrace it" rather than "fight it." For DSPD, in which the person cannot fall asleep sooner despite trying all sorts of remedies, the chronotherapy strategy is to stay up two to three hours later on successive nights, while keeping the total amount of sleep constant. Eventually, you work your way around the clock, stopping at the desired, prearranged time. In this strategy, the sleep-wake timing determines everything else, such as the time for breakfast, lunch, dinner, and activities. For example, if a person generally falls asleep at 3 a.m. and needs to sleep seven hours, then with *phase-delay chronotherapy,* he or she would stay up on the first night until 6 a.m. and set the alarm clock for 1 p.m. (seven hours of sleep), wake up and have breakfast. The second "night" he would stay up until 9 a.m. and set the alarm for 4 p.m., wake up and have breakfast. The third "night" he would stay up until noon and wake up at 7 p.m., and so on until he would reach the target of falling asleep at either 10 p.m., 11 p.m., or midnight, and then remain on that schedule—religiously, because any deviation from the schedule runs the great risk of reestablishing the ingrained "night owl" pattern. Therefore, once you achieve the desired timing, it's important to stick to a consistent schedule—going to bed and waking up at the same time every day, even on weekends. On the other hand, trying to take a major leap backward all at once (switching from going to sleep at 4 a.m. to going to sleep at 10 p.m.) rarely works.

- **Sleep in darkness.** It can help to change the sleep environment. Shift workers sometimes note that when they're forced to sleep during the day, it helps to use a sleep mask or light-blocking window shades. Also, when night shift workers go home in the morning, they should wear wraparound sunglasses to minimize any light exposure that can further alert them and interfere with falling asleep when they get home.

- **Modafinil (Provigil).** This is sometimes prescribed to combat sleepiness in shift workers. In fact, the FDA has approved the use of this medication in shift-work disorder. In a double-blind clinical trial published in 2005, researchers evaluated 209 patients with shift-work disorder and concluded that treatment with 200 mg of modafinil reduced their extreme sleepiness and caused a small improvement in their performance. They cautioned, though, that this wasn't a miracle cure, and that the treated patients still had residual sleepiness. With my extensive experience in treating narcoleptic patients and shift-work disordered patients with modafinil, I believe the study might have shown better results if the researchers had used higher doses of the medication. It is clear that many people need as much as 400–600 mg daily for substantial benefit.

"She likes to tell people, 'I get up at noon and work in my underwear,' but it's not actually true—[Ann] Coulter is rarely up before 1."

—John Cloud, quoted in "Ms. Right,"

Time magazine, April 25, 2005

TIMING MATTERS

Because of the detrimental effects on health, mood, and overall quality of life, it's important to try to combat circadian-rhythm disorders. You may be able to improve sleep timing by following general sleep-hygiene rules and using chronotherapy or bright-light therapy; if this fails, it's a good idea to consult a doctor and see what other options you can try.

Chapter 7

CONFUSIONAL AROUSALS AND THE CRUSH OF SEVERE MORNING SLEEP INERTIA

onfusional arousals are intense periods of mental confusion when someone is awakened from a deep sleep at night, or when they try to wake up in the morning. The sufferer experiences disorientation to time and place, mental dullness, slowed speech, and delayed reactions to questions or requests. Those with this disorder can even become violent or aggressive, especially during forced awakenings.[1] These episodes can last minutes to several hours, and often those who experience them have no memory of the event.

Behaviors can be either simple or complex and semi-purposeful or non-goal directed. Most confusional arousals are pretty simple: a person sits up in bed and looks around with a bewildered stare, may mumble a bit, and then resumes lying down and sleeping. Or there is rolling and thrashing around in bed in an irritable manner, ripping off one's bedclothes while uttering nasty comments at whoever is around, and not seeming to be "with it" much at all. Sometimes there is shouting, kicking, and otherwise warding off anyone coming too close. Long, disjointed talking, or loud, menacing or eerie shouting occurs while the person is lying still or moving around. The sufferer is truly out of sorts, at times acting like a bear forced out of hibernation. Interactions with others mostly involve confusion, irritability, aggression or violence, and hardly any meaningful dialogue. There can be futile promises about getting out of bed soon, and resistance to anyone trying to help out. Being "drunk on sleep" can in fact be the most accurate descriptor of confusional arousals, and is the basis for the official term "sleep drunkenness" originally coined in German. Inappropriate, vigorous, and harmful sexual activity (accompanied by sexual sounds and words) with oneself or a bed partner can also appear spontaneously during confusional arousals. Tooth-grinding (bruxism) and repetitive chewing can also occur. Because of the almost universal amnesia accompanying confusional arousals, it is the bed partner who points out the behaviors.

Confusional arousals are fairly common among children (17 percent) and also in adults under 35 years of age (4 percent).[2] The childhood form of the disorder is typically harmless, and the episodes usually happen less frequently after age five. In children, the episodes may appear quite bizarre and frightening to parents, with a child appearing to "stare right through" the parent. The child may become more agitated when parents try to console her or him.

Genetics are the most important factors in this condition—you're more likely to have confusional arousals if your parents do, too—and there's no gender difference in the likelihood of its happening. Other predisposing factors include rotating shift work, night shift work, sleep deprivation, stress, psychiatric disorders, as well as many of the other sleep disorders discussed in this book.[3] You're more likely to have a confusional arousal if you drank alcohol or used drugs that night, or were forcibly awakened.

One of the officially recognized variants of confusional arousals is *severe morning sleep inertia* (or "sleep drunkenness").[3] There can be serious complications from long-standing, severe morning sleep inertia, as Sarah's poignant story illustrates.

Sarah's Story

As a young child, I remember once waking up so terrified that nothing could make me feel safe, not even my parents, who were trying so hard to comfort me. My whole body shook and thrashed. I remember my mom yelling for my dad to get a washcloth to put between my teeth so I wouldn't bite off my tongue. They couldn't get me fully awake, no matter what they tried. I remember only this one instance; my mom tells me this happened quite often.

Even as a kid, I would sleep really late on weekends and in the summers. The only reason I didn't during the school year is because of my amazing parents. If I could change anything in my life, it would be to change what I put my poor parents through.

On school days, they could not give up—they had to get me to school. My mother would start waking me up long before she should have had to. She'd start by coming into my room, sitting on the edge of my bed and saying my name over and over: "Sarah, Sarah, Sarah . . ." She had to say my name a good 30 times before I would even realize she was there. I might have groaned or moved, but I was still out cold. This would go on for a while, and she would speak louder and louder. Sometimes I could hear her doing this, but it was as if I were in a coma. It was exhausting for me to try so hard to respond to her, so by the time I was able to say anything, I would yell: "I AM UP!"

Now, if any normal person yelled like that, I don't think there's any way they would be able to go back to sleep, but five seconds later, I was out like a light. Right back into that deep sleep, and she'd have to start all over again. "Sarah, Sarah, Sarah . . ." I don't think anyone has had their name said to them as many times as I've had mine said to me.

When she finally would get me out of bed, the whole house—and I'm sure the whole neighborhood—knew it. I'd fly out of bed, yell at her that I was awake, storm around the room, throw things. I hated her, I hated mornings, I hated my life. I wanted to die. I would have her convinced that I was awake and going to stay up, so she would leave. But the second she was gone, I was back in bed and back to sleep. My mom talks about how she would sit at the top of the stairs crying because of what she had to go through. She would then start all over.

I missed so much school, even in grade school. I knew more tricks than any other kid. But in reality, I wasn't pretending I was sick—I really was sick. My back hurt, I'd have a headache, my head would be stuffed up, and my whole body felt like I'd been hit by a bus. I felt this way for hours after I actually woke up.

When I was a teenager, my first job was as a waitress. My shift began at 11 a.m. I knew it wouldn't work. I was late every day. Not only
that, but people would ask me for more coffee or whatever, and I
wouldn't remember to bring it to them until they'd asked two or
three times. I wasn't fully awake until 2 p.m. or later. Obviously, I
failed as a waitress. I was told that I would fail at everything I ever
tried to do because I couldn't get my lazy butt out of bed.

I ended up quitting school and going to an alternative school
where I had to go only once a week and pick up my work so I could
work independently at home. I got pregnant at 17, and I promised
my parents that I would move out and raise the baby on my own. At
first when I was still living at home, I slept in my mother's room with
the baby so she could wake me up to feed and change him. I wouldn't
hear him cry. It must have killed her to watch us move out, knowing
that I wouldn't wake up for him.

I went to bed at about 9 p.m. and woke up around 9 a.m. with my
baby. I always bragged about what a good baby he was, how he slept
through the night at such a young age. But I'm sure he woke up and
cried for long periods at a time and I just slept through it. I do remember hearing him many times and not being able to get to him. I'd think
about this when I finally would get up with him the next day and see
the happiest baby in the world. Hungry and soaked to the bone, but
happy nonetheless. I know he was just thankful that his mom finally
got up for him. It kills me to think of what he went through all of
those long, lonely nights crying for someone to come help him, change
him, feed him, but most of all just be there for him and love him. But
there was no one, until I met Shane, who would become my husband. He would do everything in his power to wake me up before he
went to work every day. As soon as the door shut, I was back to sleep.

I lived a life of shame. I would lie to everyone I had to deal with
on a daily basis. Everything I lied about was because of sleep. My
mother finally made me give her a key to my apartment after she had
time and again pounded on my door, hearing my baby crying and
my never waking up for any of it.

I got my GED and went to pet-grooming school. I would wake up
at noon or later and call the school, giving them some excuse as to
why I wasn't there. I ended up going part-time in the evenings, so it

took me twice as long to finish school. I got pregnant again, so now I had another baby who was going to be neglected when Shane wasn't there to take care of him.

Shane said I was a good person and mother when I was awake, but could turn into Satan when I was waking up. We separated shortly after we were married for a year. The boys were one and three years old at the time. I think this was the toughest time of all dealing with my sleep problem, or as everyone else put it, my "laziness."

The house was very childproof, even though my kids got into many things. Cleaning supplies, knives, scissors, and other dangerous things were out of reach, out of sight, or locked up. Every day, I tried to make everything safer for the boys. But they would write on the walls, the carpet, tub, toilet, each other, and me with permanent marker. I'd sleep through it. They'd jump on my bed, cry, yell, beg me to get up and feed them. My three-year-old decided to shave his face one day. He cut himself. I lied to people about how it happened.

One time there was a bottle of spilled milk in the oven. My three-year-old said he was trying to feed the baby. Thank God the oven knobs were hidden. Another time, a lady from Social Services was standing in my upstairs bedroom doorway yelling my name to wake me up. The three-year-old had let her in. She was there checking into the financial assistance I had asked for, making sure my husband didn't still live there, but I lived in fear all the time that one day they'd come for my kids. It got to the point where I put a lock on the outside of the kids' bedroom door so they wouldn't get hurt in the rest of the house. They'd be stuck in their room from the time they went to bed at 9 p.m. until sometime in the afternoon the next day. It was real hard to get them to sleep at night, and at the time, I couldn't figure out why. They knew they wouldn't see me for a good 12 hours once they gave in to sleep.

My bed was like a drug. I would do anything to get back into it if I was forced to leave it. I didn't care who got in the way, who got hurt in the process, or who I had become during those episodes. When I was about 12, my parents had tried to get help for me, but the doctor couldn't figure it out, so nothing was ever done about it. I don't remember ever waking up in the middle of the night. When I was 21, another doctor told me to drink eight glasses of water before I went to bed. It didn't work—I wet the bed at the age of 21.

One of my friends told me that I needed to get help. She didn't sugarcoat anything. I was going to lose everything if something didn't change. I would never have guessed in my wildest dreams that my life would change just by opening the Yellow Pages and looking up the word *sleep.*

Evaluation at the Sleep Center and Response to Treatment

By the time Sarah came for her initial consultation with me, she had completed our center's standard, comprehensive questionnaire that covered sleep-wake patterns and all other aspects of sleep and symptoms of its disorders. It also covers caffeine and alcohol use, medical history, psychiatric history (including chemical dependency), and medications taken (prescription and over-the-counter).

At our initial consultation, which Sarah's mother also attended, I focused intently on the "functional consequences" of her complaints—that is, the extent to which her health and her daily life were adversely affected by her sleep problems. After determining that Sarah had major sleep issues that were severely affecting her life, I arranged for her to spend two consecutive nights at our hospital's sleep laboratory for polysomnographic (PSG) monitoring. She was also continuously videotaped in a time-synchronized fashion with the PSG monitoring throughout the night. She voluntarily underwent a urine toxicology screen, which was negative. (Sarah was not taking any medication during her sleep-lab studies.)

On the day between the two overnight sleep studies, Sarah was also tested formally by PSG monitoring in the sleep lab in a multiple sleep latency test (MSLT). This meant that she would go to bed every two hours for 20 minutes with the lights out to determine how quickly she would fall asleep, on average, during five consecutive nap opportunities. This measurement is called *sleep latency*—the amount of time it takes from the moment a person lies in bed to the time the person falls asleep. People with disorders of excessive daytime sleepiness (such as narcolepsy) generally fall asleep on average in less than five minutes (and no more than eight minutes) during these five MSLT naps. The borderline range for excessive daytime sleepiness is an average sleep latency of five to eight minutes.

It's important to note that the beginning of the first nap in MSLT comes two hours after a *spontaneous* morning awakening from the previous night's sleep. In Sarah's case, it allowed her two hours to overcome the mental sluggishness and other symptoms characteristic of severe morning sleep inertia. Furthermore, in the sleep lab, she had none of the major demands that existed in her daily life, such as caring for an infant child.

Sarah did not demonstrate any underlying sleep disorders during her two overnight studies and her MSLT. There was no evidence for nocturnal seizures that could have rendered her "wiped out" the next morning. Nor did she have obstructive sleep apnea or any other form of sleep-disordered breathing, which would have disrupted her sleep and lowered the level of oxygen in her blood, resulting in morning sleep inertia. She did not have restless legs or other abnormal behaviors during sleep. Finally, she did not have excessive numbers of repeated arousals from sleep, also known as sleep disruption. Her nighttime sleep was deep, long, and clean.

Her overnight studies showed that there was nothing wrong with her sleep except that it was lasting too long. It took Sarah an average of more than 15 minutes to fall asleep during nap opportunities, so she did not have a disorder of excessive daytime sleepiness, such as narcolepsy. She also completed extensive psychological testing, which did not indicate any major psychiatric disorder or psychopathology—except some tendency for depression. However, in a person with severe daily problems related to sleep issues, it would be surprising *not* to detect at least some depression.

And so our extensive testing was able to exclude many diagnostic possibilities for her sleep-related complaint. We could then focus on her major sleep problem: severe morning sleep inertia. Sarah's problem lay in the inability to transition from sleep to wakefulness in a full and timely manner. In its most extreme form (which we did not witness in the sleep lab but which her mother had repeatedly witnessed at home), she could not transition at all from sleep and would be "totally out of it" with agitated behaviors.

Therefore, the challenge for me regarding treatment was to help Sarah rapidly transition from sleep to wakefulness each morning, so she could get on with her day and meet her responsibilities and obligations. As it turns out, Sarah was the second patient to come to our sleep center with this problem.

The first patient was a 30-year-old woman who was referred to me by her psychiatrist, who coincidentally was my best friend. That woman was in danger of losing her job on account of frequently arriving late to work (by one to three hours) because of her severe morning sleep inertia. This long-standing problem had been resistant to the use of multiple alarm clocks placed away from her bed, having friends telephone her repeatedly in the morning, and other measures.

Since there was a long wait before this woman could see me for our initial consultation, and because she was in imminent danger of losing her job, her psychiatrist chose to treat her presumptively. He decided that his patient should take sustained-release methylphenidate (i.e., long-acting Ritalin-SR) immediately before she fell asleep each night, with the rationale being that this preparation of stimulant medication would not prolong her falling asleep but would allow her to awaken and become alert much more rapidly the following morning. And this novel treatment quickly saved the day for the woman.

When I finally evaluated her, after we studied her in the sleep lab overnight (after she had temporarily stopped taking the methylphenidate) and the next day with an MSLT, we excluded other diagnostic possibilities for her complaint, and were then able to diagnose her with severe morning sleep inertia. After we gave her a slight dose increase of Ritalin-SR (from 30 mg to 40 mg at bedtime), her symptoms completely disappeared, and she has been continuously employed for years now. She's rarely late to work. Also, despite taking a stimulant medication at bedtime, she nevertheless has slept well through the night without interruption (most people with sleep inertia tend to be deep sleepers, and also the sustained-release preparation of stimulant medication doesn't "jolt" the body system). In the morning, she has been able to respond promptly to a single alarm clock.

Given the great success of this unusual but quite rational therapy with our first patient with severe morning sleep inertia, I decided to offer this treatment to Sarah, and she also responded magnificently.

Sarah's Happy Ending

I will never forget the first night on medication. I was so excited to get out of bed to see if it actually worked, but in the back of my mind, I had a lot of doubt. The first night was the happiest night of my en-

tire life. During that night, I woke up four or five times for no reason. Nothing woke me up; I just woke up. The first couple of times,
I was so happy that I woke Shane to tell him. He would go back to
sleep and I'd lie there and smile myself back to sleep.

That next morning was a lot different. I woke up to Shane getting
ready for work. And then the emotions kicked in. I felt confused,
alone, scared, bored, happy, and maybe a little shocked. I didn't know
how the boys would react to me being there as soon as they woke up.
I woke up to a clean house. I realized that I had never sat in the morning sun in my house before. It was quiet. I didn't know what to do
with myself. It was too early to call my mom and tell her the good
news. The boys slept late, so all I could do was sit and watch the minutes pass on the clock.

I actually thought I wasn't going to like this. How did people wake
up at that time and stay sane? What did they do to keep themselves
busy?

The boys got up and I did something I hadn't done in a long time:
I made them breakfast. Time passed, and it didn't take long for the
bad feelings to go away. Now I like to wake up early because there
isn't enough time in a day.

Today I live a normal life. I am able to wake up at the first buzz of
the alarm clock, and I have the volume turned way down low. I am
able to wake up for my kids if they have bad dreams, and I'm able to
get to work on time. I am now a morning person and go to bed usually no later than 10 or 11. I wake up at about 7 a.m., put my shoes
on right away, and can't sit still until I go to bed at night.

VARIETIES OF CONFUSIONAL AROUSALS

People do some pretty strange things when they're in the grips of a confusional arousal. Sometimes they're laughable things, and sometimes they're
downright dangerous to the person or to anyone around him or her.

One of our patients earned himself the nickname Santa Claus because
of his particular variety of confusional arousal—he would jump out of
bed, wave his arm, and repeatedly shout, "Ho! Ho! Ho!" Then he'd go
right back to sleep. He never had any idea in the morning that he'd done

such a thing. His confusional arousals were triggered by obstructive sleep apnea.

Another man who came to see me with his wife told me that he compulsively scratched his anal area in his sleep. He would scratch himself until he drew blood, which was very painful. He had resorted to wearing gloves at night, and even putting tape around the end of the gloves so he couldn't take them off. When this type of activity is reported, the first thought of most doctors is that it concerns intestinal parasites (such as pinworms)—sometimes parasite activity is felt most strongly at night. But the patient had already undergone a complete medical evaluation. All tests for parasites, gastrointestinal disorders, and dermatological disorders were negative, which is why he was referred to our sleep lab.

We observed him overnight in the sleep lab and discovered that he had confusional arousals where he'd sit up, look around, then lie back down. And, yes, he did scratch his bottom. He never put his finger into his anus (as he would occasionally do at home), but he would compulsively scratch his whole perianal region.

I prescribed bedtime clonazepam, a benzodiazepine medication that is commonly used to treat confusional arousals, sleepwalking, and sleep terrors in adults, and it substantially reduced the frequency of his perianal scratching, as reported at our follow-up visit. His wife came to every visit, and she agreed that there was a beneficial response to treatment, which translated into her being less bothered during her sleep. (It was distinctly unpleasant for her to hear the scratching at night, and know exactly where he was scratching himself; it had interfered somewhat with their bedtime intimacy.) However, he still wanted to wear gloves taped at the wrist on some nights, as extra insurance.

There are certain strange, repetitive behaviors that a person performs only in the context of sleep that can become the main or exclusive focus of the confusional-arousal behavior. Nocturnal scratching is one that isn't yet classified in the ICSD, but I hope it will be included in the next edition as another official variant of confusional arousals. There are mounting case studies in the sleep literature and in the dermatology literature describing people who repeatedly scratch themselves during sleep, often to the point of bleeding and scabbing.

My wife, Andrea, has suffered from nocturnal, sleep-related scratching throughout most of her life, and often has awakened with scratch marks and bleeding on her arms, and occasionally on her face. One night, however, it wasn't her own scratching that woke her up; it was mine! I had been scratching her back and shoulder aggressively while I was sleeping. She then awakened me and emphasized that I was not being affectionate with her at all and that she was not amused by my antics. So nocturnal scratching can involve scratching either oneself or one's bed partner during sleep, and the scratching can be quite vigorous.

Another type of confusional arousal is hair-pulling. There's a psychiatric disorder called *trichotillomania* that describes people who compulsively pull their hair out, one strand at a time or in clumps, during waking hours. But that's not the same problem as pulling hair out during sleep. One woman who came to me for an evaluation actually had patches of baldness to show for her strange nightly behavior. She had a history of sleepwalking and confusional arousals, and was going through a stressful period in her life. She practiced stress-reduction techniques, and after about six months some of the stressors subsided on their own, which resolved the nocturnal hair-pulling.

Another woman was reported to have pulled out her eyelashes and eyebrows exclusively during sleep for 16 years (as observed by her family), and the first time she sought help for this problem was at a sleep center—two months before her wedding! There was no history of other parasomnia, nor any history of psychiatric disorder. A video-PSG overnight sleep-lab study confirmed that she rubbed her eyebrows in stage 1 NREM sleep, and then pulled her eyelashes in delta NREM sleep, but she was so deeply asleep that she did all her hair-pulling without arousing. She was treated with imipramine (a tricyclic antidepressant), and during the first four months of therapy there was visible, sustained growth of her eyebrows and eyelashes.

During a talk I gave at the Upper Midwest Sleep Society a few years ago, I mentioned in passing that nocturnal nail-biting can be another type of confusional arousal. A physician in the audience said, "I feel so reassured that you mentioned this in the context of confusional arousals, because I have this, and I've been wondering all these years, *What does it mean?* Am I regressing to childhood? Is it something psychological?" I think this worry is what keeps a lot of sufferers from seeking help. They jump to the conclusion that they're neurotic or otherwise mentally ill, when in fact it may simply be a confusional arousal from delta sleep. A sleep condition, nothing more or less.

My wife, who never bites her nails when she's awake, has periodically bitten her nails off in her sleep for most of her life. Often, but not always, she is in a semi-asleep, semi-awake state when it happens, and she says, "I have irrational rationalizing for why I need to bite my nails off." She has chipped her teeth from her nocturnal nail-biting, requiring visits to her dentist to have them filed down.

She has applied "bad-tasting stuff" to her nails, but still would bite them during sleep. She has applied fake acrylic nails and various other artificial nails that survived intact while she went white-water rafting, but she still bit through these nails in her sleep. When she would wear gloves to bed at night, often she would bite through them while asleep in order to get to her nails. On other nights, despite falling asleep wearing gloves that were securely wrapped with tape, by morning the gloves would be lost in the sheets and her nails would be bitten down. So nail-biting during sleep can be a "hard-core" confusional-arousal problem.

Thumb-sucking can be yet another confusional-arousal behavior that results in callus formation. A number of years ago, the *Wall Street Journal* ran a story on how up to 10 percent of adults still suck their thumbs, but my recollection is that there was no mention of nocturnal (sleep-related) thumb-suckers. We can confidently surmise that a fair share of the 10 percent of adults who still suck their thumbs will do so in their sleep and not just while awake, with others sucking their thumbs only during sleep.

I've had another confusional-arousal experience, besides the time I scratched my wife in bed. While visiting friends in Massachusetts in 1998, Andrea and I were asleep the night of our arrival when I suddenly sat up, yanked the pillow out from under her head, shoved it between my legs, and resumed lying down and sleeping—with Andrea awakening with a rude startle. I had no memory of it, but she sure did! She told me later that that moment was very abrupt and strange—almost violent.

Shortly thereafter on that same night, I was at it again in my sleep, awakening her with elaborate, incessant sleep-talking that was so intelligible that at first she engaged in ongoing conversation with me—until she realized that I was "fast asleep and unconscious" (which amazed her). At the same time, while I continued to sleep-talk, she saw me repeatedly flip straight up from side to side without changing my place in the bed. This lasted several minutes, until she became concerned that my bouncing up and down in bed would cause her to bounce off the bed, and so she went to another room, where she remained for the rest of the night.

I did have some risk factors for confusional arousals on that particular night: sleep deprivation (staying up late the previous night, packing my suitcases), a long day of traveling, the excitement of seeing old friends, drinking alcohol late at night, and then sleeping in a new environment. I had no further sleep "episodes" for the remainder of our ten-day visit on Martha's Vineyard.

When these sorts of behaviors happen occasionally, it doesn't mean you need to rush off to a doctor or work yourself up with worry. It's only if the behaviors become a pattern, disrupt your life, or endanger your safety that you need to seek help.

CONFUSIONAL AROUSALS IN LITERATURE

In the novel *Tess of the d'Urbervilles* (originally published in 1891), author Thomas Hardy provides a harrowing look at confusional arousals. The main character, Tess, has been rebuffing the advances of Alec since they met, but he waits until she's in a vulnerable state—sleep—to force himself on her. Tess had five major risk factors for a confusional arousal: her young age, severe sleep deprivation, physical exhaustion from walking many miles that day, acute stress from being confronted by a jealous female who wanted to fight with her earlier in the evening, and tremendous ongoing emotional stress—her family and Alec were pressuring her to marry a man she didn't love. When Tess collapses on the ground in the forest and almost immediately is in the midst of a confusional arousal, Alec rapes her, and because of the compromised, profoundly sleepy state she's in (her "confused surrender," in the words of Hardy), she doesn't resist.

Sleepsex

Another official variant of confusional arousals, just recently identified, is *sleepsex,* also known as *sexsomnia* or *atypical sexual behaviors of sleep.* During sleepsex, a person may do or say things that are very out of character for his or her normal sexual behavior, and this often has major interpersonal and clinical consequences. These sleep-related abnormal sexual behaviors usually occur during confusional arousals but also can occur with sleepwalking. The set of abnormal sexual behaviors during disordered arousals includes prolonged or violent masturbation, sexual molestation and assaults (of minors and adults), initiation of sexual intercourse irrespective of the menstrual status of the bed partner (unlike during waking intercourse for those individuals), and loud sexual vocalizations during sleep—followed by morning amnesia. I will discuss sleepsex more fully in a later chapter.

NOCTURNAL SEIZURES

Confusional arousals and nocturnal seizures have some similarities, but there's an important difference.

Nocturnal seizures may be prominent events or have no manifestations other than a very simple, trivial-seeming behavior, such as rubbing an eye or touching a particular body part. The seizure behavior is almost always *stereotypic:* that is, it always occurs in the same sequence. A person turns his head to the right, then grabs his left arm, then grunts. Or he always rubs his left eye with his right hand. The behavior itself isn't the important part; the exact repetitive sequence is (what French neurologists for centuries have called "the march of signs and symptoms"). Epilepsy certainly is not limited to the foaming-at-the-mouth, jerking-of-the-limbs grand mal seizure that people often envision. It can present itself in a very subtle manner during wakefulness or sleep, which is why it may be underdiagnosed. It can easily be mistaken for a confusional arousal. And to make matters more complicated, a nocturnal seizure can sometimes trigger a confusional arousal.

TREATMENTS FOR CONFUSIONAL AROUSALS[4]

We now know that sleep deprivation is the single greatest precipitating factor for confusional arousals, with alcohol use, medications, shift work, stress, sleep disorders (such as sleep apnea or narcolepsy) and mental disorders (particularly anxiety and depression) being other known risk factors. The first step for anyone who's experiencing confusional arousals is to make sure the person is getting enough sleep, on a regular schedule, and that the sleep is not interrupted by problems such as OSA. The amount and timing of alcohol use and medication administration need to be considered.

Stress-reduction techniques such as yoga, meditation, progressive muscle relaxation, or biofeedback may help.

Benzodiazepines such as clonazepam are the medications used most often to treat severe confusional arousal disorders. Tricyclic antidepressants (often in low doses) may also be used, particularly in children. For those with severe morning sleep inertia, the long-acting form of the stimulant Ritalin (methylphenidate) and/or the long-acting form of the energizing antidepressant Wellbutrin (bupropion) SR or XL, taken immediately before falling asleep, has proven to be effective on a long-term basis and usually without adverse effects.[5]

Chapter 8

SLEEPWALKING

The Twilight State

A mother came to see me once with her teenage son because he kept urinating in the wrong places while sleepwalking, most recently in the refrigerator, but also previously in his closet, on top of the toilet seat with the lid down, or on carpets. This is funny stuff, but I didn't laugh. It's no fun cleaning up after a sleepwalker who needs to use the bathroom but pees in the wrong places. That is why his mother brought him to our sleep center. Enough was enough, even though the pediatrician told them that he would almost certainly outgrow this problem (not necessarily true at all, we now know).

Sleepwalking, or *somnambulism,* is pretty easy to define—it's when someone gets out of bed and walks around or does other activities while still asleep. It's a disorder of arousal, and it produces an altered state of consciousness where people are able to perform complex actions without awareness of what they're doing.

Up to 17 percent of children have sleepwalking episodes. It usually starts before puberty, peaking at age 11 or 12. Most people think that kids will just grow out of sleepwalking—and this is often the case, but it's also more prevalent in adults than most of us realize. At least four percent of adults are prone to sleepwalking.

The sleepwalking events themselves are quite unpredictable. Some people are content to just shuffle around the room a bit before returning to bed. Others do complex or dangerous things—like walking outside in the snow in a nightgown, rearranging furniture, driving across town, or cooking a meal. Interestingly, men have a higher injury rate during their sleepwalking episodes.

Most of the time, sleepwalking begins in deep delta sleep (NREM stages 3 and 4) during the first third of sleep, though it can also begin during stage 2 NREM sleep.

Sleepwalkers typically have little or no memory of the episodes the following day (with some notable exceptions), and it's difficult to awaken them during an episode. There's a chance that startling sleepwalkers can make them aggressive and even violent, but the risk inherent in waking a sleepwalker is generally minimal.

One of the important things to note about sleepwalking is that it generally has no psychological significance. Sleepwalking is an involuntary act and does not indicate that a person is mentally ill or has any sort of underlying psychological disorder.

Jamie's Story

I started sleepwalking when I was a teenager, maybe 14 or 15 years old. From the beginning, I nearly always woke after having been at it for a while, and one of the things I did most often was unplug electric gadgetry from the wall sockets: the fan, the DustBuster, the stereo, the lamps. Funny thing, though: I never touched the alarm clock. I seemed to know not to make myself late for work by disabling the alarm. In that regard, sleepwalking has always seemed akin to lucid dreaming—which I also do. Part of me is aware, or at least my eyes are open, and I think I'm perfectly safe to walk around without hurting myself. I can see and hear, but it's as if the computer processing is disconnected from the motherboard.

The peak of my sleepwalking episodes was between the ages of 14 and 20, during all of which time I lived in one-level apartments. When I moved to a three-story town house, I found that I would not venture down the stairs in that state. Often I would wake up standing at the top of the stairs, looking down. Once, I woke up folding shirts and putting them back into my bureau drawer. I had emptied them out and was in mid-refold—gave me quite a chuckle.

Mostly, these things make me laugh, although there were two incidents that left me uneasy. Once, I woke up on the closet floor, wrapped shroud-like, head-to-toe, in the bedsheet. Not a good look if you're still alive. And the most recent one was unlike any of the others, as I have no recollection and the evidence of it was suspicious. In the morning, I woke as usual but noticed dirt and scraps of mulch in the bed where my feet had been. My shoes were outside—very odd—and later, my perpetually drunk neighbor asked me if my sister was visiting, because he thought he had seen her out on the deck in the wee hours of the morning. Needless to say, she wasn't there, but it seems perhaps I was. I didn't like that one at all.

NOCTURNAL TREASURE HUNTING

Anthony, a technology consultant, says, "Last night I vaguely remember dreaming about personal digital assistants (PDAs) and, more specifically, frustratingly fumbling with an expansion card trying to get it in the slot of my PDA. I also remember being cranky a few times when rolling over in bed as something hard jabbed me in the back or arm. Well, this morning when I pulled the covers back and got up, I found I'd collected in my sleep my PDA, an expansion card, a Walkman, and a set of headphones. I'm not usually a sleepwalker, but I guess last night I must have gone for a wander, picked up all this gear, and then returned to bed and tried to insert the expansion card into the PDA slot."

CHILDREN AND SLEEPWALKING

In childhood, sleepwalking can usually be considered a normal developmental sleep phenomenon (an occasional quirk of sleep), apart from those cases having clinical consequences, such as injury. So it's not usually necessary to treat a child's sleepwalking, though some precautions are in order.

One of the lucky things about sleepwalking in children is that it generally occurs during the first hour or two after a child has gone to sleep, which means that parents are usually still awake. If a child wanders out of bed, a parent can see it and direct the child back to bed before any harm is done.

All children's bedrooms should be made childproof, meaning that there are no sharp objects, matches or lighters, or fragile items in the child's reach. A child prone to sleepwalking should not sleep in a bunk bed, and windows should be locked. Electrical cords should also be out of the way to reduce the possibility of tripping.

In most cases, sleepwalking does not become a problem unless it goes on late into the sleep cycle—or late enough that the parents are already asleep. This opens up the possibility that the child could fall down stairs, open the front door of the house and walk outside, or do something else dangerous. In this case, there should be an alarm on the child's bedroom door, and the house doors should be locked in a manner that's difficult for the child to unlock (out of reach, double-locked, etc.). Otherwise, just being watchful parents is enough.

I was thirsty again, so I lighted a candle and went to the table on which my water-bottle was. I lifted it up and tilted it over my glass, but nothing came out. It was empty! It was completely empty! At first I could not understand it at all; then suddenly I was seized by such a terrible feeling that I had to sit down, or rather fall into a chair! Then I sprang up with a bound to look about me; then I sat down again, overcome by astonishment and fear, in front of the transparent crystal bottle! I looked at it with fixed eyes, trying to solve the puzzle, and my hands trembled! Somebody had drunk the water, but who? I? I without any doubt. It could surely only be I? In that case I was a somnambulist—was living, without knowing it, that double, mysterious life which makes us doubt whether there are not two beings in us—whether a strange, unknowable, and invisible being does not, during our moments of mental and physical torpor, animate the inert body, forcing it to a more willing obedience than it yields to ourselves.

—From *Le Horla* by Guy de Maupassant, 1887

RISK FACTORS

Genetic factors play a role in 65 percent of sleepwalkers. In fact, family history of sleepwalking is the greatest predictor of whether a child will develop sleepwalking tendencies. Also, the incidence of sleepwalking increases in relation to the number of affected parents. If one parent is affected by sleepwalking, the chances are 45 percent that one of the children will also be affected by sleepwalking. If both parents sleepwalk, then there is a 60 percent probability that one of the children will also sleepwalk. This shows that sleepwalking usually has a physiological cause, rather than a psychiatric cause.

Sleep deprivation is the major trigger for sleepwalking episodes, so sufficient sleep and a regular sleep-wake pattern can help a great deal. Many other precipitating factors have been identified, such as hyperthyroidism (increased thyroid hormone levels), migraine headaches, head injury, seizure disorder, encephalitis (brain inflamation from infection), and stroke. Eating spicy or high-fat foods close to bedtime can also be triggers.

Obstructive sleep apnea and hypopnea, and other forms of sleep-disordered breathing, which I described in previous chapters, are becoming increasingly recognized triggers of sleepwalking in both children and adults. Once the sleep-breathing problem is controlled, then the sleepwalking typically stops.

Fever can bring on sleepwalking episodes in children who are predisposed to the condition. Travel, jet lag, sleeping in unfamiliar surroundings, emotional stress, and the premenstrual period can also trigger sleepwalking, along with use or misuse of alcohol, caffeine, and some medications (particularly psychiatric medications such as lithium and amitriptyline, and also zolpidem [Ambien]).

In children, the urge to urinate (an internal trigger) may cause an abrupt arousal leading to a sleepwalking episode, so it's best to make sure they go to the bathroom before bed and limit eating and drinking before bedtime.

In our sleep clinic, we've also seen two cases of injurious sleepwalking with sleep terrors (screaming while terrified) that occurred only when the women were in their premenstrual states, which points to possible hormonal triggers of disordered arousals.

Ambien and Sleepwalking

Although it's apparently rare, there are some cases of zolpidem (Ambien) triggering sleepwalking episodes while being used properly to treat insomnia. Among those who've had this side effect, sleepwalking stopped when the medication was discontinued.

A recent study done on the link between zolpidem and impaired-driving traffic arrests indicates that even though alcohol and illegal drugs are sometimes involved, the behavior of the zolpidem drivers is generally more bizarre than other people who've had too much to drink or are driving under the influence of drugs. They may drive in the wrong direction, slam into light poles or parked cars, and seem "oblivious to the arresting officers."[1]

In many such cases, either the person took too much zolpidem or drank alcohol in conjunction with the medication—which the label warns against. Though the cases were not described as sleep-driving (and we can't know conclusively how many are and how many aren't), I'd suggest that cases of sleep-driving are more common than most of us realize. In many arrests where zolpidem was found in the blood, the driver claimed to have no memory of driving or the accidents they'd been involved in. And in some cases, the driver has been clad in a nightgown or pajamas, further suggesting a lack of awareness when he or she walked outside and stepped into the car.

In a less-mortally-dangerous and more amusing scenario, doctors recently described the case of a woman who managed to "sleep-e-mail." She was taking zolpidem and had been sleepwalking for about six months. One day, she got a phone call from a friend of hers to say that she was accepting her dinner invitation. The woman had no memory of inviting her friend to dinner.

She checked her sent e-mail, and sure enough, about an hour and a half after she had gone to bed the previous night, she had written three very strange, badly formatted e-mails to her friend, asking her to come over for dinner and drinks. The woman had no memory of writing them. She must have walked to the next room, turned on the computer, connected to the Internet, signed on with her password, and written and sent the messages—all while in a sleepwalking state apparently induced by zolpidem.

So writing e-mails is another manifestation of this phenomenon in which modern tools of communication at our disposal are used "under the influence."

CHARACTERISTICS
OF SLEEPWALKING

Sleepwalking episodes usually share certain features:

- **No warning.** There is usually no physiological forewarning that a sleepwalking episode is about to occur, that is, there is no increase in muscle tone or twitching, no increase in heart rate or breathing immediately prior to the episode, no substantial change in the brain waves—just quiet, uneventful slow-wave sleep, until the "alarm rings," the sudden central nervous system arousal triggers heart-rate acceleration and the sleepwalking begins.
- **Glassy-eyed stare/blank expression.** Parents with children who sleepwalk often report that the child's eyes spook them—sometimes they describe it as the child seeming to "look right through" them. During an episode, the pupils are dilated.
- **Disorientation upon awakening.** The person may be confused and bewildered, and have no idea how he or she got out of bed. The disorientation goes away within a few minutes.
- **Meaningless talk.** Talking while sleepwalking is common, and it's usually nonsensical rambling.
- **Amnesia related to the event.** Sleepwalking children don't usually remember anything about what they said or did during their episodes. For adults, it can vary considerably: They may remember nothing at all, have a foggy recollection, or recall the events fairly well, sometimes with associated dreaming.

Sometimes the sleepwalker will immediately jump up and start walking at the onset of an arousal, or can bolt from the bed and begin running. There can be agitated, belligerent, or violent behavior, again, especially in adult men. Often, sleepwalking involves inappropriate behaviors, such as urinating in a wastebasket, moving furniture around haphazardly, or climbing out a window.

Sleepwalking is often combined with sleep-talking in children and adults, and in adults, sleepwalking is often combined with sleep terrors, which involve intense, sudden screaming and agitated behaviors. A person who is sleepwalking or having a sleep terror sometimes may be in a dream-like state—he could be "seeing" attackers or other threats to his safety and may respond accordingly.

One of my patients, a 25-year-old married woman, was sleepwalking in the kitchen and dreaming about an intruder, so she threw a jug of water across the room to hit him. Another time, she dreamed she was pulling pins out of her daughter's mouth, and awakened to find her fingers in the girl's mouth—pulling out the imaginary pins. But her most troubling incident was on Christmas Eve, when she placed her infant daughter under the Christmas tree during a sleepwalking episode. She had been dreaming of putting a doll under the tree as a present for the girl. We were able to get her disorder completely under control with clonazepam.

Another patient came to us with only one complaint about her sleep-walking—she would always overfeed her cat. The animal had now become obese with medical complications, and she was quite concerned about controlling her sleepwalking. Since she never remembered what she did while asleep, it took her a long time to link her cat's obesity to her sleepwalking. However, she was now convinced that she was a "sleep-feeder" (her expression).

Gentlewoman: Lo you, here she comes! This is her very guise; and, upon my life, fast asleep. Observe her; stand close.

Doctor: How came she by that light?

Gentlewoman: Why, it stood by her: she has light by her continually; 'tis her command.

Doctor: You see, her eyes are open.

Gentlewoman: Ay, but their sense is shut.

—From *Macbeth* by William Shakespeare

WAKE THEM OR LET THEM SLEEP?

Ideally, if you are trying to help sleepwalkers you just want to direct them back to bed rather than trying to awaken them. However, it's more dangerous *not* to awaken sleepwalkers if they continue to engage in harmful activity than it is to awaken them. There's a high risk of injury with sleepwalking: tripping, falling, cutting oneself, and so on.

It's best not to startle sleepwalkers by shouting at or shaking them, if this can be avoided, because you do run the risk of startling and frightening the person and possibly having someone get hurt in the process (the person may trip, or take a swing while in a confused state). Waking them up in a gentle manner is preferable. But it is a myth that you shouldn't wake a sleepwalker because you might "scare them to death." This myth comes from a belief among some primitive cultures that a person's soul left his or her body during sleep, so if you woke up a sleepwalker, the soul would be lost.

TREATMENTS FOR SLEEPWALKING

There are several treatments that can eliminate or lessen sleepwalking episodes, assuming that any problem with sleep deprivation or irregular sleep hours has been corrected. An effective strategy for parents of children who are having periods of frequent sleepwalking and/or sleep terrors involves waking the child shortly before the usual time that an episode appears— typically within an hour of falling asleep—for several consecutive nights.

Self-Hypnosis[2]

One of the more popular treatments for sleepwalking—if it persists despite proper sleep habits and is not posing any harm to the person—is self-hypnosis. Hypnosis is a state of altered consciousness where you're very receptive to suggestions and orders because your critical thinking is low and your attention is highly focused. In a practical sense, hypnosis involves a trancelike state of focal concentration during relaxed wakefulness. Also, "hypnosis" *means* "self-hypnosis," since very few people can be hypnotized involuntarily. Hypnosis is not sleep, per se, although it can lead to sleep if the person so wishes.

In a study done at the sleep center, we reviewed the cases of 27 adults who had come to us seeking help with noninjurious sleepwalking and sleep terrors, and who had been taught a self-hypnosis technique developed by my former colleague and lead author of that report, Thomas D. Hurwitz, M.D., to be practiced nightly for several minutes. The technique involves having the patient sit reclined in a recliner chair, keeping his or her legs uncrossed so as not to create any pressure points. The clinician sits nearby and has the patient roll his or her eyes while slowly shutting the eyelids, and then has the patient relax the eyes while keeping them shut.

Then the clinician guides the patient through a progressive-relaxation breathing technique by having the person take a deep breath through the mouth, hold the air in to create some pressure in the chest (which causes tension), and then slowly exhale. This causes a feeling of relaxation, a welcomed contrast to the tension of holding the breath. The mouth breathing encourages awareness of cool and pleasant air being inhaled. This breathing sequence, as tension and relaxation is generated over and over again with a progressively slower respiratory rate, allows for relaxation, a necessary part of the hypnotized state.

Various muscle groups are then identified for contraction and relaxation (beginning with any muscle that feels tense at that moment), as part of the progressive-muscle-relaxation sequence. Then the patient visualizes himself or herself lying quietly in bed asleep throughout the night—but if at any time he or she sits up or puts a foot on the floor, then that would trigger the switch for full and immediate awakening, alertness, and awareness of the surrounding environment. Putting a foot on the floor promptly triggers an awakening. In other words, no "twilight state" of mixed sleep and wakefulness—a hallmark of sleepwalking—is allowed.

After reviewing these cases, we found that 74 percent of these patients reported much or very much improvement in their symptoms when they used bedtime self-hypnosis techniques that we taught them during office visits. A person doesn't need lengthy lessons in hypnotherapy to make this effective; most of our patients learned the techniques in just one or two sessions, then practiced at home. One patient learned the techniques from a psychologist who gave her the hypnotic suggestion that whenever she was about to start sleepwalking, a flashing blue light would awaken her and prevent her from sleepwalking. She would see the psychologist every six months to review and reinforce this technique, which has been very successful.

Sleepwalkers can also use relaxation tapes at night or practice yoga. It's best to learn self-hypnosis techniques from a trained professional. Hypnosis is proven effective in both adults and children.

Benzodiazepines

Benzodiazepines are usually very effective in controlling sleepwalking, and can be safely prescribed to those with chronic, injurious, or otherwise worrisome cases. The mechanism of action appears to be "raising the arousal threshold," meaning that a much stronger central nervous system stimulus or external stimulus (loud sound, light) is required to trigger sleepwalking. Clonazepam (Klonopin) is frequently used, but various other benzodiazepines can also be effective. There is a point to emphasize: Because sleepwalking (and sleep terrors) can emerge as soon as 15 minutes after falling asleep, and usually within the first hour or two of sleep, it is important to take the medication about 30 to 75 minutes before falling asleep, so the medication is given sufficient time to be absorbed and delivered to its therapeutic locations in the brain.

The patient should discuss with the doctor the expected duration of treatment, and a plan to eventually attempt to stop the medication. At least two hours should separate taking a benzodiazepine at bedtime from modest alcohol use earlier in the evening, or else potentially major adverse effects could occur.

Clonazepam (Klonopin—U.S.; Rivotril— Other Countries)

The way clonazepam acts varies, depending on the parasomnia being treated, which indicates how versatile it is as a therapeutic sleep agent. Its abuse potential is quite low—people don't "get high" on clonazepam (unless they abuse it with methadone). In fact, taken in excess it can cause a "downer" feeling. Clonazepam is a nontoxic agent, posing negligible risks to liver, kidney, lung, and other organs; frequent or special blood tests are not required; there are no particular concerns about long-term treatment;

and it is compatible with many other classes of medication. Once a proper therapeutic dose is found, then that same dose will usually continue to be the effective dose during long-term treatment.

At least 1 percent of my patients treated with clonazepam at bedtime for a sleep disorder have experienced substantial side effects early in the course of treatment, such as morning over-sedation, memory problems, unsteady gait, depressed mood, "dark personality," hair loss, reduced libido, erectile dysfunction, or acid reflux. These symptoms are completely reversible upon lowering the dose or discontinuing the medication. In my experience, most patients with disturbing clonazepam side effects can be safely and effectively switched to alprazolam (Xanax), another benzodiazepine that is about half as potent as clonazepam.

Environmental Remedies

For those who live alone or whose sleepwalking episodes may be dangerous, a door alarm can offer a measure of protection. There are several types of door alarms that attach to a bedroom door and sound off if the door is opened. The idea is that the sound of the alarm should fully awaken the person and interrupt the sleepwalking episode. If the sleepwalker lives with others, the alarm can alert someone to wake up and go check on the sleepwalker.

All sleepwalkers should try to make sure their bedrooms are on the first floor of a home, apartment building, or college dorm, and plan ahead to request the first floor when staying in hotels.

Sleepwalkers should not have weapons (loaded gun, knives) accessible in the bedroom or in any other room of the house. Weapons should be locked with the key hidden from the sleepwalker.

TREATING UNDERLYING CONDITIONS

Sleep doctors from Stanford examined the records of adult sleepwalkers and found a surprising result: 99 out of 101 of their sleepwalking patients had either obstructive sleep apnea (OSA) or a milder form of sleep-disordered breathing (with reduced airflow) called *upper airway resistance syndrome* (UARS), with UARS being the more frequent problem.[3] So they began a study: 50 chronic sleepwalkers (average age: 24 years) underwent PSG testing, received treatment, and had a follow-up one year later. The researchers also looked for other sleep disorders and psychiatric disorders, testing the theory that treating associated or underlying disorders had a better chance for success than treating the sleepwalking itself.

Eight of the 50 sleepwalkers who were evaluated had depressive or anxiety disorders. Instead of treating the sleepwalking, the researchers chose to treat the psychiatric disorder first. This didn't help the sleepwalking. Five had sleepwalking only (no other sleep disorder or psychological disorder) and were treated with benzodiazepine medication at bedtime. All five dropped out of the study, citing limited relief—the sleepwalking persisted among all of them, even though the frequency of events decreased. (However, the dose of the medication used was not mentioned in the report, so it is possible that a higher dose would have been effective; also, the sample size of five patients was very small for any valid conclusions.)

The remaining sleepwalking subjects were all diagnosed with sleep-disordered breathing and no psychological disorder. The first option presented to all of them was nasal CPAP therapy. Those who were compliant with this therapy had a complete absence of sleepwalking along with control of their sleep-disordered breathing at the six-month mark. Those who weren't compliant with CPAP were offered surgery. Many chose not to have the surgery, but among those who did, all reported no more sleepwalking episodes one year after the surgical treatment.

So, in this study, properly treating the breathing disorder was the solution in every case. However, this doesn't hold true in all cases. Referral bias to the "tertiary sleep clinic" at Stanford may have played a role, since the experience at our center and at other centers is that sleepwalkers with sleep-disordered breathing may continue sleepwalking despite effective CPAP treatment, and so they need separate treatment of the sleepwalking. It is a provocative study, though, showing that treating the underlying condition may be the right answer for controlling sleepwalking in some, or even many, people.

Jan's Story

Although I was diagnosed with a severe sleepwalking disorder as an adult, I was probably sleepwalking since the time I could walk. When

I was four years old I would get up, turn on all the lights in the house, open all the doors, and go out for a walk. My mother told me about it because they talked about tying me to my bed. Sometimes I would run through the house and yell, "Charge! The white horses are coming!"

Or I would wake somebody up. Sometimes I would wake up in strange places, in closets, back porches, and basements. As I got older, it got progressively worse—I was trying to climb out of second-floor windows. It was like I was living a dream I was having. I was really experiencing that dream. I also did a lot of talking in my sleep.

Most of my dreams are people trying to chase me, running from someone trying to kill me, or trying to rescue someone. However, I would also take my dogs for walks with me, and that was calm, so I don't think there was terror all the time. When I was younger especially, I would go outside and go for a walk down the block. My mother would come after me after she realized I was up and out. But I never remembered anything. I didn't wake up.

My younger brother was at home with me in my teens, and he was always afraid I was going to kill him along with his friends because of the weird look on my face that people get when they're sleepwalking. He said my eyes would be just bulged out and glazed over, wide open. My daughter says the same thing.

When I'd go on trips or family outings, my brothers would play with me in my sleep to get me to go, to get me active, so everybody thought that was a joke. I was always worried when I'd go on trips. When I went on school trips or with other friends, I'd try to stop that from happening. A lot of times, I'd spend the night awake to make sure I didn't sleepwalk. Now I never stay in a room with somebody else because of the things I do.

When I was a teenager, my dad had made a comment about my room being messy before I went to sleep, and when I woke up in the morning, I had two beds in the room. They were both in the center of the room at that time; one was made (that was the one I went to bed in), the other had no sheets on it or anything. I had unloaded my closets, all my shoes and books and everything, onto that bed, and I was sleeping under them when I woke up. All the other furniture was in the center of the room.

There were many times I'd wake up in different parts of the house. Sometimes I would be totally naked and not know why. And when I looked down the stairs to figure out where I was, there was an image of a shadow in the doorway or window, and I thought it was a man who was out trying to catch me, and I saw his shadow in the second-floor window, too, so I just felt kind of imprisoned until I could slide down the steps to go to my room and then get dressed for bed.

Of course, I was always exhausted in the morning. I felt like I'd been run over by a Mack truck. I never felt good. I would be active almost nightly, and I'd always wake up exhausted. I could sit down to watch a show and fall asleep, go to a movie that's really exciting and sleep through it. About the time that other people should be getting up and going to work, I could finally go to sleep for a while. I was leaving some fantastic excuses at work, coming up with all kinds of good things.

Shortly after I was divorced from my first husband, my daughter was sleeping in my room. She was about nine or ten. She woke up to me sitting on the edge of the bed looking out the window. At bedtime, I had the shades drawn all the way, but when she woke up I had the shades up halfway so I could see the cemetery across the road, and I was talking to people in the cemetery. My daughter said I was asking them who fed them, what they ate, and if they ever got lonely. It was a regular conversation, she said. I would ask the question and then wait for a period of time like I was getting answers. When she woke up and heard this, she asked me if I was awake and I turned around and looked at her with my glazed-over eyes and said, "Yes."

She left the room that night and never slept with me again. But I didn't find out until my daughter was a teenager that she was really scared of me. It made me feel really bad that she would have these fears. She still doesn't sleep in the house with me now. Usually when I would wake up from my sleep disorder, people would laugh about it and I would feel kind of foolish because I didn't know what the heck went on. Kind of like being the brunt of a joke that was weird. Sometimes I would even laugh.

Jan provided this additional anecdote at a recent clinic visit: During one episode in her childhood, when she was growing up on the family farm, she opened the front door to the house late at night and called in all 32 dogs that lived on the farm! That created quite a scene in the household, as everybody was awakened from deep sleep by 32 dogs running around, yelping. Jan was definitely not a popular person in that home for quite a while afterward.

Evaluation at the Sleep Center and Response to Treatment

When I first saw Jan, she was 48 years old and had three major sleep complaints. First, her sleepwalking, screaming at night, and other strange nocturnal activity had been an ongoing problem for more than four decades, and she was quite fed up with it. Moreover, she was disturbed to hear from her husband that she would regularly swear loudly, become angry, and have a "spacey appearance" while wandering around at night. It was as if she had become an alien being on the move after falling asleep. Second, she never felt rested in the morning and attributed this miserable daytime feeling to her sleepwalking and being active most nights—in other words, not sleeping soundly. Third, she would snap her jaws tight and grind her teeth during sleep, and her dentist had noticed that her enamel had worn down. So all of her, even down to her teeth, was worn out from her parasomnias.

What was also interesting for me to hear from Jan was the vivid dreaming that she often experienced while suddenly sitting up in bed and engaging in her sleepwalking ventures. By that time in my career, I was already quite aware that many adults with sleepwalking (and night terrors) had dreaming—including elaborate dreaming—with their non-REM sleep parasomnias, and that the textbooks were wrong in saying that dreaming did not occur with these conditions.

At one point during our interview, things got interesting in a different way when Jan told me she wished she could get her sleepwalking controlled so she could go on a cruise. She began by saying how she was prone to having "weird dreams" of being in a "normal situation such as a public place" and then suddenly finding herself topless and greatly embarrassed with her arms covering up her naked breasts—and this could be the way she awakened in her home. She mentioned how she would be horrified to have that happen to her on a cruise ship—away from her cabin. This reminded me of another case I had recently seen.

Jan leaned forward with curiosity as I explained that recently I had evaluated another woman with long-standing sleepwalking, and one night while on a cruise ship in the Caribbean, she had wandered outside her cabin while sleepwalking and was not aware that her cabin door would automatically lock shut behind her, and she had no key. Well, she eventually woke up quite a ways down the hall from her cabin, clad in a see-through nightgown while staring directly at a smartly dressed young officer working the night shift, who wore a smirk on his face. In her words, "I was mortified." And so I wholeheartedly agreed with Jan that she should definitely have her sleepwalking under complete control before boarding a cruise ship.

It was also clear to me that Jan's condition was a well-entrenched physiological disorder of sleep, not psychological, since her sleepwalking had begun at such an early age—four years or earlier, virtually all her walking life—and the repertoire of her nocturnal activity was so typical of the nonsensical things that sleepwalkers do. For example, she would move furniture around, empty the closets, deposit her clothes around the house in a haphazard manner, wake up in the morning far away from her bed, and sometimes bruise herself from bumping into things. When talking to the dead people buried in the cemetery, she did not show any negative emotions and did not seem distressed in any way. Instead, she was having a mundane type of conversation (despite the bizarre context) during an episode of sleepwalking.

In the sleep lab, Jan demonstrated classic behavior for sleepwalkers, with many confusional arousals and beginnings of sleepwalking episodes emerging from NREM sleep, when she would raise her head, mumble and talk nonsensically, chew and grind her teeth (called *sleep bruxism*), and attempt to get out of bed. Also, 12 times per hour, she would spontaneously awaken briefly, with or without behaviors, before falling back to sleep rapidly. No other sleep abnormalities were found. The next day, in the Multiple Sleep Latency Test, she did not demonstrate objective excessive sleepiness, since her average time to fall asleep during five nap opportunities was nearly 14 minutes. Therefore, her complaint of feeling "wiped out" during the day was tiredness (and not sleepiness) that was presumably an adverse consequence of excessive physical activity at night, including sleepwalking, along with too many awakenings.

When I subsequently met with Jan to go over the results of her sleep-lab studies, I explained that she had, unequivocally, lifelong sleepwalking, and we agreed with a chuckle that she had never outgrown it and probably never would. Therefore, she needed some type of clinical intervention. We acknowledged that she continued to have sleepwalking events with excessive frequency and bruised herself too much from knocking into furniture, and that she felt too tired upon awakening in the morning, which carried over into most of the day.

I then mentioned our sleep center's extensive research, which showed how certain medications taken at bedtime were very effective and safe for controlling problematic sleepwalking. After considerable discussion about this form of treatment, Jan agreed to take clonazepam (Klonopin), a benzodiazepine medication that works by "increasing the arousal threshold," as I previously discussed.

Jan responded promptly and very well to bedtime clonazepam therapy, and she noticed that not only did it completely stop her sleepwalking, but it also resulted in her feeling much more rested in the morning and throughout the day, allowing her to accomplish more in her life and to feel better all around. For 12 years now, it has kept her disorder under control, and she has not needed to increase the dose of the clonazepam. She takes one or two (and occasionally three) tablets of the lowest dose strength, depending on certain circumstances (such as travel).

And so I have only needed to see Jan for yearly checkup visits, at which time I go over her sleep, the clonazepam therapy, any medical update (including medications), and an overall discussion about her life and any issues that I should be aware of. There is even some time at our visits for her to show me her current photographs, as she is an avid and talented amateur photographer.

It is important to note that Jan is still prone to sitting up at night, and talking or shouting on occasion, which indicates that her inner drive for disordered arousals with sleepwalking and related activity is still alive and well, and needs nightly control with clonazepam. However, we both have repeatedly agreed that the primary goal of therapy is to control the problematic parasomnia behaviors, and not to extinguish every single nocturnal behavior, or else we would run the risk of ramping up the dose of clonazepam to the point of causing side effects and toxicity. This calls attention to how a physician and his or her patient need to collaborate in identifying exactly what needs treatment, what needs to be controlled—what we as physicians call "the target symptoms" meriting therapy.

We also discussed her daughter's fear and reaction about how she would never sleep in a room with Jan again because of the strange behavior while sleepwalking. I encouraged Jan to have honest family discussions about her sleepwalking, to explain that current research has justified its classification as a "disorder of arousal" from NREM sleep. In other words, it is a physiological disorder of sleep, which nevertheless can produce psychological symptoms such as screaming or talking to dead people in the cemetery.

It is easy to see how family members can misinterpret and become frightened by what is going on with their loved one at night—they may even fear that they might eventually develop the same thing, which they could consider to be a form of mental illness, with frightening implications. Therefore, it should be explained that while sleepwalking seems to have a psychological basis, it is actually a result of physiological abnormalities, that is essentially the brain misfiring during sleep, which leads to disordered arousals and complex behaviors. Otherwise, as that child grows up, she or he may have growing fears of becoming "psycho" during sleep just like Mom. That could translate into eventual sleep phobia or other potentially serious psychological problems. At any appropriate later stage of development, the genetic basis of sleepwalking can be presented and discussed, and hopefully any fears allayed.

Part of the family discussion should therefore be to allow the children and others to openly express their concerns and fears about what they have witnessed at night, and at times the injuries they themselves may have sustained by being in the way of an agitated sleepwalker, or from being struck while trying to stop a sleepwalking event or while redirecting a sleepwalking parent back to bed. They should not take it personally. For sleepwalkers with younger children, there may be the fear that the parent will run away one night and never come back. Also, since sleepwalking can in fact sometimes have a genetic basis, the sooner such frank discussions take place, the better prepared, in the proper way, will the offspring be in case they too develop sleepwalking or night terrors.

Jan's Happy Ending

I've really come to terms with the sleepwalking disorder. If I take the medication, I do not get up and walk. Rarely, I may sit up and yell at somebody or call them into the room, but I don't walk or run through the house yelling "Charge! The white horses are coming!" any longer. The only time I think of the sleep disorder as a problem is when I travel and may have to share a room with an acquaintance who doesn't know about my sleep problems.

LIVING ALONE: HOW DO THEY KNOW IF THE TREATMENT IS WORKING?

Adults or young adults who live alone and therefore don't have any observers to tell them about their nocturnal behaviors face an interesting challenge: How do they determine if the treatment is working?

I advise them to get a newspaper and assemble its sheets around their bed. If the newspaper is not disturbed in the morning, they can assume that they didn't sleepwalk during the night. Of course, if they wake up during the night to go to the bathroom or get a snack, they have to make sure the papers are placed carefully back on the floor before going back to sleep.

Another alternative is to put a piece of paper in the door of the bedroom. If that piece of paper is on the floor in the morning, they know they opened the door.

GOING TO COLLEGE

Every July and August, I know I'm going to get phone calls from parents of soon-to-be college students, wanting to make urgent appointments to have their children evaluated for sleep disorders. What brings young adults in to see us most often at this age are sleepwalking and night terrors.

It never really bothered them before, but now that they're going away to college, they're suddenly very worried about what their roommates and dorm-mates will think if they wake up in the middle of the night and do something weird—like wandering naked through the hallway and shouting something about killer clams.

The other popular time for young adults to see us is when they're getting into a serious relationship and are worried about what their partners will think.

I expect those calls from parents in July, and they're shocked to hear there's a two-month waiting list to be evaluated in the sleep clinic. The student is going off to school at the end of August or beginning of September—now what? So consider this a little friendly advice: The time to think about treating a sleep disorder is when they're filling out applications to college, not when they're packing to leave. Leave enough time to get an appointment, to have testing done, and to evaluate treatments and make sure they're working.

Joni's Story

When I was general manager of a hotel, I fielded a call from a top client. One of their new employees was attending a two-week training class and had a sleepwalking problem. He was a male in his late twenties or early thirties, who slept in his tightie whities. I checked with the night auditor, and yes, the man had been roaming around the hotel and lobby at night in his underwear, much to the embarrassment of some female guests. One evening, he engaged another guest in a lengthy, if disjointed, conversation in the lobby. Another night, he was wandering, dazed, throughout the parking lot. Needless to say, some guests were concerned about what appeared to be someone mentally unstable.

The client (his boss's boss) was mostly concerned that he was embarrassed when he realized he was wandering (sometimes in front of new coworkers) in his underwear at night. I was more worried about his safety.

The client actually wanted me to padlock the trainee's room, from the outside, when he went to bed every night. I explained that I couldn't do that . . . If there was a fire or other emergency, he would die, and his survivors would own the hotel (after the lawsuit). Disappointed, she then asked me to provide someone to "watch" him throughout the night. She said it was important that we not wake him up, but wanted someone to follow him around. Unfortunately, I didn't really have the staff to provide night care to adults!

We settled on a compromise. She bought him some pajamas, and he dead-bolted and chained his door every night (which you should do in a hotel anyway), plus put his desk chair under the doorknob. That proved complicated enough that he got out of the room only once during the remainder of the two weeks.

WHEN SLEEPWALKING IS DEADLY

Although it's unusual, it's entirely possible for a person to commit a violent act while sleepwalking. Based on case studies and sleep research, we know that it's possible for a person to commit assault, suicide, or even murder while in a sleep state. What's tricky, of course, is knowing exactly when sleepwalking really is to blame. Those who've killed while asleep haven't exactly been hooked up to polysomnograph machinery in sleep labs while they did it. It's impossible to definitively know afterward if a person was sleepwalking (or experiencing another parasomnia) at any particular point in time, but we can take into account the person's sleep, medical, and family history, and we can study a person in a sleep lab after a violent attack to see if parasomnias are able to be documented.

Behavior done while sleepwalking is an example of *automatism:* an act that is not controllable by the person and is done without conscious thought. Sleep is by definition not a conscious state. Across the country, automatism is an allowable defense against crimes, including murder, in many states.

Not Guilty of Murder[4]

In May 1989, Kenneth Parks of Toronto, Canada, drove 14 miles from his home to his in-laws' home and beat his mother-in-law with a tire iron and stabbed her to death, while cutting the tendons of his hand (and later demonstrating no pain—*somnambulistic dissociative analgesia*—in the presence of police, as they later testified in court). He also beat his father-in-law until the man lost consciousness. He claimed to have been asleep the whole time, and after hearing testimony from mental-health professionals and sleep specialists, a jury believed it. They could find no motive for the attack—the 23-year-old man reportedly got along well with his in-laws and had no history of violence.

He did, however, have an enormous sleep deprivation that had built up due to job stress and gambling problems. He was physically exhausted from playing rugby on a hot, humid day. He had a history of sleepwalking, including one episode of nearly going out a window. His family was replete with parasomniacs, including a grandfather who would fry eggs and then leave them on the stove while he was engaged in sleepwalking. Kenneth was acquitted of murder and attempted murder, and the Canadian Supreme Court upheld the acquittals.

Most people who've never had complex parasomnias find stories like this extremely hard to believe—how could someone do all these things and not wake up? But it's amazing what we can do in our sleep. People have scaled buildings, gone swimming with alligators, and boarded trains while sleepwalking. People have injured themselves badly and still remained asleep. And some episodes can last for an hour or more.

Our sleep center's policy is not to give expert opinions on specific cases on behalf of either the prosecution or defense. Instead, we provide pertinent scientific and clinical information on sleepwalking and other aspects of parasomnias to any attorney or judge who contacts us. When we're asked if certain behaviors are possible while sleepwalking, the answer is almost always yes. But there is always speculation involved when we're asked to determine as experts whether or not a person was sleepwalking when performing the behaviors in question.

> In all of us, even in good men, there is a lawless wild-beast nature which peers out in sleep.
>
> —From *The Republic* by Plato

Pseudo-Suicide

What if a suicide isn't really a suicide after all?

That was the troubling question we confronted when the parents of a college student from Iowa contacted the director of our sleep center, Mark W. Mahowald, M.D., to ask us for a formal review of the official cause of his death. The local paper had called it a suicide, and that's what was on his death certificate—but the parents were sure the young man had been sleepwalking at the time.

A review of his medical history and the circumstances of his death painted a pretty clear picture. First, he had a history of frequent, complex sleepwalking—and a remarkable family history of the disorder, as well (his father, two brothers, a sister, and a cousin all had sleepwalking disorders). There were no signs of depression or drug use at the time of this "suicide," and he had been sleep-deprived prior to the event. At 4:30 a.m. on a February morning, he had walked outside in 30-degree weather wearing only boxer shorts (which his parents told us was completely out of character for his modest wakeful personality), ran onto a highway, and was hit by a semitrailer truck. He sustained head injuries and was pronounced dead on arrival at a local hospital.

After a formal review of his records and the statements from several friends and family members, we were able to recommend to the Iowa state medical examiner that the cause of death listed on his death certificate be revised to "accidental death due to sleepwalking." The record was thus properly changed.

Several times since then, the same question has come up: Family and friends want to know if a "suicide" was really an accident resulting from a parasomnia. A close friend of a police officer contacted us in 2000 to tell us about his friend's death. The man had shot himself in the mouth sometime between 4 and 8 a.m., and it didn't make any sense to his friends and family: He had short- and long-term plans for the future, no signs of depression, no identifiable reasons to explain suicide. He'd had a normal conversation with his fiancée just before he went to bed, and had asked a fellow police officer to wake him up for work the next morning.

There was a strong case for sleepwalking. He had a long history of sleepwalking, involving guns in particular. He had been discharged from the U.S. Marines due to sleepwalking, apparently because he tried to get into a locked gun case while he was asleep. Although he never hurt anyone while sleepwalking, he had once pointed a gun at his mother, and had punched a hole in walls.

We were not able to conclusively determine whether this man's death was accidental or purposeful, but it certainly left us suspicious that his parasomnia sleep disorder played an important role in his death.

There is compelling evidence of how sleepwalkers come very close to situations that could have falsely been termed suicides. When a fireman in London responded to an emergency call from a passerby who had spotted a body on top of a crane, he thought he was dealing with a suicidal person ready to jump to her death. Instead, he was astonished to find a 15-year-old girl curled up fast asleep on the arm of the 130-foot structure.[5]

He was terrified to wake the girl, for fear that he'd startle her and she'd fall, but he was able to grab a mobile phone from her pocket and find the phone number for her family. Her parents told him that the girl was a frequent sleepwalker. She was brought down in a hydraulic lift unharmed.

What does this amazing sleepwalking story tell us? Certainly, this is an example of non-agitated yet extremely dangerous sleepwalking behavior in which the girl was able to negotiate her environment in exquisite fashion, demonstrating motor dexterity and intact hand-eye coordination—all with suspended judgment in her sleep mode as she put herself at mortal risk by climbing the crane, and then curled up in a classic sleeping posture while her brain and mind remained asleep. Therefore, injurious sleeping or otherwise life-threatening sleepwalking does not necessarily involve clumsiness, agitation, or violence.

In another case reported in 2005, a 39-year-old German man climbed out a window, shimmied up a drainpipe, and walked across his roof while sleepwalking.[6] When he awoke, he lost his balance and fell 20 feet to the ground. Police say he was saved only because the lawn was soggy from rain.

Malingering

Because of its potential to remove responsibility ("I was asleep! I didn't know what I was doing!"), sleepwalking has its share of fakers. These fakers pretend to have been sleepwalking, when in fact they were perfectly awake and aware of their actions—intentional behavior that often results in harmful actions against another person. This is known as *malingering.*

Malingering is not considered a psychiatric illness. A person might pretend to have a sleepwalking disorder to avoid implication in a crime, or to excuse violence against a bed partner, or to shake off (legal) responsibility for any number of bad behaviors that may have happened during normal sleeping hours.

Many people associate malingering with people who claim false symptoms to gain money from insurance companies. While the person may have legitimately been in an accident, the symptoms described do not exist or are exaggerated. Similarly, a malingerer may genuinely have had sleepwalking episodes in the past and discovered what a handy excuse they could be—therefore pretending to have further episodes whenever he or she felt like doing something wrong.

That's exactly what a jury decided happened in the case of Stephen Reitz, a commercial fisherman from California who killed his 42-year-old married lover while on a romantic weekend trip to Catalina Island in 2001.[7, 8] He was in his mid-twenties at the time, and by all accounts, the two had been getting along perfectly well earlier that day. They'd ordered pizza, walked around town, and played card games in the hotel.

Then, soon after midnight, he struck her in the head with a flower pot; beat her severely enough to cause a broken jaw, six broken ribs, and a dislocated shoulder and wrist; stabbed her with a plastic fork in the back of the knees; and stabbed her in the neck repeatedly with her pocketknife. He claimed that he awoke to find her dead and believed he was the one who had done it because the wounds to her neck looked like the sort of injuries he inflicted when he killed a shark he caught by severing its spinal cord. He also claimed to remember that he had been dreaming of confronting a male intruder.

He immediately called his parents and told them he thought he had killed his girlfriend. They told him to go for help, so he washed his hands, got dressed, and headed to a fire department, where he confessed that he believed he'd killed the woman.

In the murder trial, the defense's sole argument was that Stephen had been sleepwalking at the time of the attack. A leading sleep expert testified on the defense's behalf, noting that Stephen had a long history of sleepwalking and exhibited all the classic signs of the disorder.

That defense might have worked if not for the compelling testimony of witnesses and Stephen's own parents, who all painted a picture of a violent, unstable man. Stephen and his girlfriend had been drinking alcohol and using cocaine that night, which was a common occurrence according to friends of hers. Several witnesses testified that about six months before the woman's death, Stephen had broken through a plate-glass window to her apartment with a knife. They also noted that they'd seen bruises on her body numerous times and that she once allegedly told a friend that she knew he'd kill her one day, but he was "like a drug" to her and she couldn't seem to pull herself away from him.

The deputy district attorney never disputed that Stephen was a sleep-walker, but he did not believe Stephen was sleepwalking during the attack. He argued that Stephen's sleepwalking experiences were what enabled him to know all the right things to say to make his defense sound plausible.

The jury found Stephen guilty of first-degree murder, and he was sentenced to 26 years to life in prison. A Superior Court judge rejected the request for a new trial or a reduced sentence.

Scott Falater, a Mormon father from Arizona, similarly explained his actions after he stabbed his wife 44 times and rolled her into the pool, holding her head underwater: He also claimed to have been sleeping the whole time.[9]

A neighbor called police when he saw Scott drag his wife into the pool. In the initial questioning and throughout the trial, Scott maintained that he loved his wife, they hadn't argued, and there were none of the hallmark motives for murder—no extramarital affairs, no financial distress, no substance abuse, nothing to explain why he'd murder her.

He also had a history of sleepwalking since childhood, and had been sleep-deprived due to work stresses: He was a unit manager and product engineer struggling with a failing product line.

But perhaps the reason a Phoenix jury found Scott guilty of first-degree murder was the "cover-up" effort: Police found Scott's bloody clothing, the knife, and his boots and gloves neatly stored in a container in his Volvo. The evidence seemed to indicate that Scott had changed clothes and gone into the garage at least twice to hide his clothing and weapon while he claimed to have been sleepwalking. The neighbor testified that Scott had recognized his own dog during the attack and motioned for him to be quiet—why would Scott recognize his dog and not his wounded wife? Prosecutors also noted that his wife was not wearing her wedding ring when her body was found.

The sleepwalking defense is rarely used, and it can be quite difficult to convince a jury of its veracity. Jurors, therefore, need to be informed of the broad range of known sleepwalking behaviors (including driving a car, cooking a meal, sleepsex, and violence) when they assess the evidence in a particular criminal case that uses the sleepwalking defense.

In what is likely the first case where the sleepwalking defense was used successfully, the public clearly thought the defendant was malingering—but the jury took only a few hours to acquit him of murder. In 1845, Albert Tirrell, a married man from Massachusetts who had been previously arraigned for adultery, slit a prostitute's throat and severed her jugular vein and windpipe.[10] Then he set several fires in the brothel before fleeing to a nearby stable to try to find a horse to ride home to Weymouth. Relatives harbored him, then he escaped the state before being arrested on a ship in New Orleans.

At the trial, prosecutors argued that Albert killed his mistress, Maria, because she refused to stop being a prostitute, even though he had left his family to be with her. But Albert's lawyer insisted that if Albert had killed her at all (he raised the possibility that she had committed suicide), he did so while sleepwalking. Not only did he escape the murder charge, but he was also acquitted of arson charges at a separate trial.

WHAT SLEEPWALKING RESEARCH HAS TAUGHT US

Sleepwalking still has a few mysteries. We know that sleepwalkers have slow-wave sleep instability; they usually have arousals early in the sleep period, which almost always occur during the first or second slow-wave sleep period, even on nights in which there is no sleepwalking. What is currently unknown are the exact mechanisms that cause this instability, although genetic factors play a major role in this process, as already described in this chapter.

Also, in 2003, researchers discovered that sleepwalkers carry an increased probability of having a certain gene in the "HLA complex" related to the immune system. This is not a "sleepwalking gene"—not all sleepwalkers have it, and about 13 percent of the non-sleepwalking population has it, too—but the scientists concluded that those with the genetic marker are 3.5 times more likely to be sleepwalkers. This genetic marker is in the same complex as markers for narcolepsy and REM sleep behavior disorder. It's likely that future research will focus on the HLA complex to find out more about the genetic influences in sleep disorders, and their interaction with the immune system, which may ultimately lead to highly specific therapies in cases that warrant intervention, such as injurious sleepwalking.

Recent research has demonstrated that sleepwalkers almost always show some "sleep brain waves" (theta and delta EEG activity) during their episodes, no matter how simple or elaborate their behavioral repertoire. So, in other words, parts of their brains really are asleep, even when they're roaming through a hotel or climbing down a drainpipe.

We're learning more today about the possible triggers of sleepwalking, with the role of sleep-disordered breathing, including obstructive sleep apnea, gaining increased attention in sleep-related medical publications, in both children and adults. Controlling any sleep breathing problem can then lead to control of sleepwalking, although this is certainly not always the case. We are also identifying certain medications that can set off sleepwalking, which may in the future help us understand the neurochemical mechanisms promoting sleepwalking.

Chapter 9

ABNORMAL SLEEPSEX

Unknowing Cruel Intimacy

S leepsex is one of the recently defined parasomnias that is made up of a spectrum of abnormal sexual behaviors and experiences.[1, 2, 3, 4, 5]

Sleepsex (or sexsomnia) can run the gamut of sexual behaviors. Most often, it includes sexual vocalizations; "sleepsex-talking"; masturbating (often violent or vigorous), touching, grabbing, or fondling the bed partner's genitalia; or sometimes, complete intercourse. Oral and anal sex can also occur during sleepsex, but they are less common.

Sometimes, as you might expect, the sufferer doesn't consider the disorder a problem. A single person who discovers that she masturbates during sleep or a man whose wife finds it amusing that he initiates sex while sleeping may not find their behaviors troublesome. Some partners enjoy the experience of sleepsex. Women often note that their boyfriends or husbands are "different" sexually when they're asleep—usually more aggressive and commanding, but sometimes more gentle and focused on pleasing.

However, more often, the behavior is unwanted and can be downright dangerous to the person with sexsomnia, the bed partner, or anyone who happens to be nearby when the person falls asleep.

For example, in one case, an Australian woman with a steady partner learned that she'd been having affairs in her sleep.[6] Her partner knew that she was prone to sleepwalking, but after he found condoms strewn around the house, he became suspicious—and she had no explanation. One night, he woke to find her missing and tracked her down . . . having sex with a stranger away from their home.

A colleague at another sleep clinic recently treated a happily married couple with two teenage daughters who was planning a three-week family trip to Europe in which they would all sleep in the same room in order to save money. Only there was a big problem: The mother had developed frequent and loud sleepsex-talking several years previously, so she and her husband came for help to stop her lewd sleep-talking before the trip. After her evaluation in the sleep lab, she was diagnosed with an NREM parasomnia, and bedtime clonazepam therapy substantially controlled the sexsomnia before the family trip. Just in time.

While these accounts may sound hard to believe, sleepsex is a real disorder that can have serious consequences for many people.

Serenity's Story

I've been with the same man for more than three years, and my experiences with his sleepsex behaviors have been very frequent. They've ranged from silly to scary and everywhere in between.

First of all, he has this pervasive habit of dry-humping my behind in his sleep. It happens most when he goes to bed intoxicated or (oddly enough) at times when our sex life is more active than usual. My favorite position in which to sleep is spooning with my back against his stomach, but unfortunately, this position seems to encourage his behavior. While definitely asleep, he'll grab my hips and really just grind himself into me, often with noises to accompany his actions.

As you can imagine, in the early stages of our relationship, I found this pretty humorous and sometimes even erotic—in a "He can't even keep his hands off me while he's sleeping!" sort of flattered manner. Now, more than three years later with a bit of the sexual glow worn off from our initial "hump like bunnies" infatuation, it's just plain annoying. He wakes me up in this fashion at least one night a week, and I have to literally retreat to the farthest edge of the bed and face him with my knees drawn up to keep him off me so I can get some sleep.

He's exhibited a few other sleeping sexual behaviors as well.

Twice, while I was reading in bed while he slept, he's rolled over, pulled his penis out, and started slapping it against my leg. Both times I asked him what he was doing, and he (while still apparently asleep) calmly replied he didn't know, put his man-parts away, rolled over, and settled back into a normal sleep.

While those two incidents still get me giggling when I recall them, a few nights have actually been frightening. Every once in a while, my boyfriend's sleepsex behavior has become violent. I've awakened in the middle of the night to his forcefully trying to pull my underwear off me, sometimes partly in the act of inserting Tab A into Slot B, or violently groping my breasts. One night he actually ripped my underwear off, and another time he gave me a sort of rug-burn–type abrasion from my bikini-underwear straps. There have been times I've been forced to just get up and go sleep on the couch.

These behaviors, while I'm certain he doesn't intend to do them, are quite troubling to me for a variety of reasons. So far, I always wake up in time to stop him from (I hate to say it like this, but I can't think of any other appropriate term) violating me, but I often wonder what will happen if I don't.

Usually, in his sleep, he seems to respond to my orders to leave me alone, although he never seems to entirely wake up during or after these incidents, and he never remembers doing any of these things or my telling him to knock it off. I've brought it up several times during our relationship, informing him of his previous night's behaviors, whether amusing or abusing, and he never remembers any of it.

What I find perplexing about his behavior is he's such a gentle, caring man when he's awake. I suppose there's something to be said for subconscious desires in this case, but I'm not the one to puzzle them out.

CHARACTERISTICS OF SLEEPSEX

I recently searched the medical literature and found 31 documented cases of parasomnias with sexual behaviors; 80 percent were men, usually in their early thirties, who had been exhibiting sexsomnia for almost ten years, always with amnesia. Many also had been sleepwalkers beginning earlier in life, and often had family histories of parasomnias. Masturbation, fondling a bed partner, forced sexual intercourse, and sexual vocalizations and verbalizations were the most common sleepsex behaviors. Snoring during sleepsex occurred in three patients (who had sleep apnea) and in almost half the cases, agitation or assault occurred during sleepsex. Nine cases involved sleepsex with a minor, and there were legal consequences in 11 cases. Injuries such as bruised penis or breast, vaginal abrasions, lacerations, and scratch marks were common.

Almost all of the patients were diagnosed with confusional arousals and some also suffered from sleepwalking. The good news is that bedtime clonazepam was effective in 90 percent of treated cases. A few patients also had obstructive sleep apnea and snoring during sleepsex, and CPAP therapy got their problems under control.

Sleepsex experiences can be more playful or loving, or aggressive, but also "robotic." There's a lack of kissing—the "sleepsexer" may head straight for the genitals. People often rip off clothing (their own or their bed partner's). They are often more aggressive than during wakeful sex. In a *Details* magazine article, a 26-year-old man described his sleepsex episodes, which happened a few times a month: "You know how on *National Geographic* when an animal sneaks up behind a female and jumps on and pumps away? That's me."[7]

Not only is the manner often different, but the acts themselves are often things the person would not do when awake. Certain behaviors that are normally taboo or not pleasurable while awake may be performed freely when asleep. For example, a man may pay no attention to the fact that his wife is menstruating.

Sleepsex may be combined with sleepwalking: The person may leave his or her bed and enter someone else's bed, or masturbate in a different room. But more frequently, it seems, the person remains wherever he or she fell asleep.

Frequently, if a partner tries to "join in" or accepts the advances, the person will wake up and react with annoyance or anger—"What are you trying to do to me?" The person is completely unaware that it was himself or herself who initiated sex.

Some people are easy to fend off; if the partner is not in the mood for sexual activity, it's simple enough to yell or give the person a shove and the behavior will stop. Other people are nearly impossible to stop, which is why sleepsex sometimes feels like rape, even between a married couple.

Sleepsex may go on for long periods of time, and if the initiator wakes up (naturally or by force), he or she may simply roll over and go back to sleep, even if the partner wants to continue. Sometimes the person wakes up while having an orgasm.

It can be very difficult to convince people that they did something sexual during sleep, particularly if the act is out of character. Normally, they don't remember the event at all, and may only realize what's going on if they are faced with evidence (a videotape or audiotape, proof of ejaculation, waking up in the middle of an act, waking up in the nude, etc.).

The relatively small number of clinically evaluated sleepsex cases (31) reported in the medical literature probably reflects the low awareness that clinicians have about sleepsex, a condition that only received official recognition in 2005 in the ICSD-2. It is very likely that there are many other sleepsex sufferers, both initiators and recipients, who have gone undiagnosed and untreated.

SLEEPSEX IN THE COURTROOM

As you might imagine, the subject of sleepsex has caused a lot of controversy and difficult decisions in the legal world. What do you do when a person who is asleep sexually assaults someone?

In May 2005, 34-year-old Richard Anderson was accused of molesting two girls, who were ages 11 and 13 at the time of the incidents.[8] One of the girls told a school counselor, who contacted the Department of Social Services. There were two separate incidents: In the first, the man picked up the 11-year-old from a friend's house and lay down with her until they both purportedly fell asleep. The incident with the 13-year-old occurred when she wasn't feeling well and she asked him to lie next to her. In both cases, the girls awoke to him touching them inappropriately. At his trial, however, he had an unlikely supporter: the girls' mother.

During the grand-jury trial, one of the girls testified that she knew it was an accident and that *she knew he was sleeping because he was snoring.* Their mother was very unhappy that it had ever become a criminal trial and claims that the girls' words were distorted and the child-interview specialist presented false testimony. She said the girls never wanted to testify against him.

Richard pled guilty to reduced charges of assault and battery and was sentenced to three years of probation, during which he was ordered to attend sex-offender counseling and have no unsupervised contact with the girls. (The news reports never explained how he knew the girls.)

In another tricky court case in 2001, Richard Overton claimed that he had been sleepwalking when he got into bed naked with a seven-year-old girl his girlfriend was babysitting.[9] He says that when he awoke on top of the girl, he asked the girl what she was doing in his room—and she replied that he was in *her* room. He grabbed the bedspread and ran out of the room. Although he was found guilty of endangering the welfare of a child and child abuse, he appealed the ruling.

Richard's sister, girlfriend, and ex-girlfriend had all testified that they had witnessed his strange sleepwalking episodes before, sometimes finding him asleep in the bathroom, other times under a table. They stated that he often performed complex behaviors—like turning on burners on the stove—and had no recollection of these events the next day.

The prosecutor persuaded the jury that Richard was purposely endangering the children in the home by not putting locks on their doors and by sleeping naked, but the appellate court decided that this was not enough to support a child-endangerment charge. Because sleepwalking acts are not voluntary and because the trial judge failed to instruct the jury to ignore certain comments made in the prosecutor's closing argument, the convictions were overturned.

A similar case in 2005 first ended in a mistrial when a jury was deadlocked. They heard the story of Jonathan Hutchinson, a supposedly sleepwalking man who wound up in bed with his girlfriend's 14-year-old daughter at least three times.[10] He claimed to have no memory of the events, and remembers only that he awoke to the girl pushing his hand away each time. In a second trial, the defense called upon Jonathan's mother, who testified that he had been a sleepwalker since childhood, sometimes doing strange things like frying apples on the stove in his sleep. But prosecutors argued that he had told the teen afterward to keep the events a secret—which was not the action of an innocent sleepwalker, in their estimation.

The second jury found Jonathan guilty of three counts of child endangerment, but not guilty of sexual assault, and he was sentenced to five years in prison. He is appealing the ruling.

And in December 2004, a Norwegian court heard the case of a 33-year-old man who says he was asleep when he had sex with a sleeping 18-year-old at a party.[11] The man had been drinking, and claimed that he awoke when he heard her screaming for help. Two of the three judges on the panel ruled to acquit the man because they couldn't determine that he was conscious; therefore, the man was set free.

Malingering

As you can imagine and as with other parasomnias, there is the potential for people to claim they have sexsomnia as an excuse for behavior they intend to commit.

A 44-year-old charity worker from Essex, England, had a crush on a woman who repeatedly told him she wasn't interested in a relationship. But one night, she allowed him to sleep on her sofa—and around 6:30 a.m., he came into her bedroom in his underwear and asked, "I know you are not interested in me, but could I have sex with you?"[12] He claimed to have been sleepwalking when he pinned her to the mattress and assaulted her (stopping short of rape), ignoring her screams, then offered to make her tea to "calm her down." He left when it became clear she wanted to call the police. Then he tried to win her forgiveness by bringing her a number of "apology gifts," including a microwave, flowers, and a book about astrology. She finally called the police, and the man pled guilty of indecent assault in November 2003.

He was sentenced to three years of community service and required to attend a sex-offenders program.

There are some ways to tell if a person is malingering—such as a careful review of medical history and sleep complaints, reports from family and former lovers, and so on. In this case, the fact that the man seemed to have a grip on the true situation by acknowledging that she was not interested in him and asking her permission anyway, then trying to "calm her down" by making her tea, is not in line with typical sleepsex behavior.

However, conversely, it is possible that there are people in prison now who honestly have no memory of their crimes, and never had intent to commit them, because the defense sounded ludicrous—*No one is going to believe that I was asleep when I did this!* The defendant may never understand what happened, may never tell anyone what happened, or may be advised by a lawyer that the story is unbelievable.

And to take it a step further, at least one woman who later realized that she herself was suffering from sexsomnia nearly reported a man for date-raping her. She woke up with her skirt hiked up and her underwear pushed aside, and accused her date of trying to rape her in her sleep. She had no idea that she had initiated the behavior until years later, when she repeated the same behavior on other men she dated (including her husband) and a female friend.

SLEEPWALKER AS VICTIM

In a startling case from 2001, it was suggested that both the victim and the perpetrator were sleeping. A 19-year-old Danish woman with a history of sleepwalking was renting an apartment in London with her fiancé. One night, she sleepwalked into her landlord's room, where she says he raped her. He claimed that he had taken sleeping pills that night and had no memory of the woman ever being in his room. He suggested that she must have initiated sex, but the jury didn't buy it—he was convicted and sentenced to seven years in prison.[13]

NEGATIVE CONSEQUENCES OF SLEEPSEX

Aside from the potential criminal consequences, those who experience sleepsex often face difficult challenges.

- **Relationship distrust.** Partners often have a difficult time believing that the person is "really asleep" during episodes. This can cause tension, distrust, and anger. Couples at times report being on the brink of divorce because of chronic sleepsex problems.
- **Forced sex.** Even within the context of marriage, wives of men who experience sleepsex often say that they feel like they're being raped by their husbands. The husband may be very forceful and

unresponsive to protests. This problem can be even worse when experienced by a new couple, or with a woman who's been sexually abused in the past.

- **Exacerbation of existing relationship problems.** If the couple's waking sex life is unsatisfactory, sleepsex may aggravate tensions about it. A partner may get annoyed that a person masturbates during sleep but refuses sex while awake, or that sleepsex seems to be triggered by dreams about other people, or that one partner's needs aren't being met, or that the person may have sleepsex with others, or about a number of other issues.

- **Risk of being overheard.** Children, roommates, and neighbors may hear the loud vocalizations or bed movements that go along with many sleepsex episodes.

- **Problems sleeping away from home.** Those who are prone to sleepsex need to avoid sleeping in places where their behavior could cause a problem. Sleeping at friends' or relatives' houses may not be possible.

- **Difficulties with new relationships.** Some sufferers report that they would rather stay single than risk the embarrassment of having a new partner discover their behaviors. Imagine having a date sleep over for the first time and hoping you don't violently masturbate or try to initiate intercourse half an hour after you fall asleep!

- **Guilt and shame.** Many people are humiliated by their sleepsex tendencies. Some are embarrassed by what they do to others, and some are ashamed of their own sexual behaviors during sleep. Many look inward and wonder if they're repressing these aggressive tendencies during wakefulness. The acts may also be in opposition to their (or their partners') religious beliefs.

- **Pain and bruising.** After a violent episode, a person's genitals may be bruised, sore, or even bleeding.

- **Pregnancy and STDs.** It's less likely for a person to use protection during a sleepsex episode, so the risk of pregnancy and sexually transmitted diseases is increased.

- **Confused sexuality.** Though rarely reported, in some cases, a heterosexual person will attempt a homosexual act while sleeping—and presumably, the opposite can occur, too. This is most frequently cited among friends who are sleeping at each other's houses. Does it indicate "wish fulfillment"? As with most analysis of sleepsex behaviors, we can't say for certain, but there is no evidence to support this. You should not make the assumption that sleepsex behaviors are indicative of suppressed wakeful desires.

- **Doctors who don't understand.** Many doctors have never heard of sleepsex, which only makes it worse for the person who's afraid to seek help. Sufferers are often afraid that doctors won't believe them or won't take it seriously—and unfortunately, their fears are often justified. If a general practitioner isn't helpful, ask for a referral to a sleep specialist. At minimum, nearly all sufferers are embarrassed to talk about sleepsex.

HOW DO PEOPLE PERCEIVE SLEEPSEX?

Michael Mangan, Ph.D., an adjunct psychology professor at the University of New Hampshire, Durham, spent two years collecting stories and surveys from his website, www.sleepsex.org, for the book *Sleepsex: Uncovered*.[14] At the time he wrote the book, 204 people had responded to his survey and 51 people shared their personal stories in writing or by phone interview.

He attempted to discover how people perceived their sleepsex experiences, and concluded that there was a "dark side" and a "light side" to the disorder. In most cases, the behavior was unwanted, and often perceived as dangerous.

However, he writes, "In no case involving intercourse did a man complain that a female had assaulted him. What was especially disturbing to men was the feeling that they were guilty of sexual assault."

Women quoted in the book did say that men complained about the women's sleep-masturbation and sexual-vocalization habits, however. One woman reported that her boyfriend thought she was lying to him about being asleep while she masturbated, which made him angry. Her website (www.sleep-masturbation.com) indicates that they eventually broke up because of this issue. Another said that her husband seems angry when she vocalizes sexually—as if she's done something wrong or is hiding something from him.

The "light side" showed, however, that sleepsex was sometimes pleasurable and beneficial to the relationship. People reported that they felt more freedom during sleepsex, and that it even added to the trust of the relationship sometimes, enabling them to be more open with each other about fantasies.

WHOSE DISORDER IS IT?

One man says, "Occasionally when I roll over in my sleep and hug my wife, she'll seem sexually responsive. She'll take my hand and transfer it either to her breast or between her legs (or maybe I dream she does?). I'll caress her a little, judging her state of arousal. When I'm sure the feeling is sexual, sex ensues . . . then I get accused of waking her up for sex. As—in my mind, at least—it was her who initiated it, this is most perplexing! I'm not sure at which point she actually wakes up. A couple of times she has claimed to be unaware of anything until partway through sex, but I think generally she wakes up sooner rather than later. I'm sure this is triggered by dreams, but I'm unsure if they're hers or mine!"

Lynn's Story

There was a period when it was pretty common for my husband to try to initiate sex while he was asleep. We both were asleep. It was startling at first, and took me a while to figure out that he was sleeping. I remember attempting to talk to him, but getting mostly mumbles and a strange attitude, sort of seductive, aggressive, and determined . . . like He-Man, not his usual bedroom style. He had no memory of this on the mornings after. I believed that he didn't remember anything, but it also surprised me, as it wasn't a small interruption.

His usual method was to begin groping at me, sometimes fondling my breasts, but usually going straight for the gold (to put it nicely). He was aggressive about it, and persistent. If I pushed his hand off, he would come right back at me with more determination. If I tried to go along and turn it into an actual sexual encounter, he would not wake up—he would just keep grabbing and rubbing, and not in a very enticing way, rather rough and quite annoying. It was a bit disappointing sometimes, as I wouldn't have minded following through.

I finally figured out that if I just sternly commanded him to roll over and go back to sleep (even though he *was* asleep), and sometimes grabbed his hands and placed them back on his own body, he would finally stop. (Unfortunately, once I was awakened, it was very hard for me to get back to sleep.) This happened most often during his nights off of work. He worked a graveyard shift, and his sleep schedule was constantly off-kilter; we slept together only a few nights a week. And finding time for intimacy was difficult. This behavior went on a few times a month for about a year or two.

The behavior stopped after he went on a two-month sabbatical leave from work. The sleepsex seemed mostly related (in my opinion) to our faltering sex life, which was a result of his graveyard shift. He just wasn't on a good sleep schedule and wasn't getting all of his needs met, and was over-stressed as well.

Once he went back to the graveyard shift after the sabbatical, we had worked out the sex thing by providing ourselves other opportunities. Also, we worked hard to put him on a more consistent sleep routine, so he was getting closer to 7 or 8 hours at the same time every day, rather than the sporadic schedule he had before, which could range from 4 hours one day to maybe 10 or 12 on a day off.

WHAT CAUSES SLEEPSEX?

Sleepsex is a variant of confusional arousals in the preponderance of reported cases, but as we've discussed, it can occur with sleepwalking. Obstructive sleep apnea (OSA) can trigger sleepsex with snoring during confusional arousals. Treatment with continuous positive airway pressure (CPAP) should usually control the OSA and sleepsex.

Episodes of sleepsex seem to arise from some mysterious or obscure internal, hair-trigger alarm mechanism that thrusts a person out of deep delta NREM sleep and into a confused quasi-awake state. All sorts of primitive and confused behaviors are then unleashed, from bulimic eating, to aggression and violence, to sleepsex. These partial awakenings from slow-wave NREM sleep are called disordered arousals. Any number of triggers can cause these arousals. Alcohol can set off an episode, for example, or getting poor sleep three nights in a row.

WARNINGS AND PRECAUTIONS

For those who have unwanted sleepsex experiences, some precautions are in order.

Because alcohol and sleep deprivation are two major triggers, eliminating or cutting down alcohol use and getting regular and sufficient sleep are two important ways to protect against sleepsex episodes. Drug use and stress may also play a role. However, in many cases, sleepsex episodes arise spontaneously, without any known trigger.

Those who experience sleepsex should not sleep in the nude unless the only other person in the home is a bed partner who doesn't mind. Further, it's best to sleep in pajamas that are not easy to take off (i.e., not just a long shirt, but a top and bottom with buttons or other hindrances).

It's important, particularly if there are children or visitors in the house, to take safety measures, such as locking bedroom doors or using a door alarm. Those who are prone to sleepsex should never allow a child or unsuspecting friend to share their bed.

If you vocalize during sleep, you may wish to sleep with a white-noise machine (such as a fan) or radio to drown out the sounds.

Some couples opt to buy separate beds, or to sleep in different rooms, if the sleepsex episodes are too frequent or problematic.

TREATMENT OPTIONS

The benzodiazepine clonazepam (Klonopin) has been 90 percent effective in controlling sleepsex originating from parasomnias. Controlling any underlying sleep-breathing condition such as obstructive sleep apnea can completely resolve sleepsex, as well. This normally involves treatment with a CPAP machine, but could also involve a surgical procedure on the upper airway.

FUTURE SLEEPSEX PROGRESS

The forensic aspect of sleepsex is receiving quite a bit of attention recently because of high-profile cases. It's important for the public to understand that sleepsex is real before we're able to pass judgments between those who have the disorder and those who are malingering. If you check the Internet for discussions about rape cases where sleepsex has been the defense, you'll find that the vast majority of people immediately assume it's a bogus excuse for rape and that judges and juries are naïve to believe it. That's certainly possible in any particular case, but we also know that people do instigate sex—sometimes forcefully—while asleep and are unaware of their actions.

My colleagues and I have already published reports and have spoken with lawyers and judges about enhanced guidelines for cases like the ones mentioned earlier in this chapter.

Chapter 10

SLEEP TERRORS

The Night the
Ghost Got Scared

A piercing scream erupts in the darkness. A mother stumbles out of bed and rushes to her child's room. Is he sick? Is he hurt? Is there an intruder in his room?

"What's wrong?" she asks while turning on the light, but he doesn't seem to notice she's there. He just screams and sobs, flails around, eyes bulging in terror.

Grabbing his shoulders and shaking them even a little bit seems to agitate him more. And it's like he's seeing something frightful in the room that isn't there, and doesn't recognize his mother. He looks possessed and seems to be gazing right through her, which sends a shudder through her. What is going on?

"Wake up, Kyle. It's okay. Nothing's happening to you."

But nothing she says seems to get through to him. He bolts out of bed and toward the door, sweating and panting. He's much stronger than usual, and is whipped into a screaming frenzy. After about ten torturous minutes where Mom's sure he's going to have a heart attack and she's going to have a breakdown, she's able to lead him back into bed, and he quietly returns to sleep.

In the morning, he remembers nothing about this. He's tired, though. And so is Mom, who had a hard time returning to sleep after seeing her son go through this torment. *This didn't look like a normal nightmare,* she thinks—and she's right. Her son Kyle had a sleep terror the night before.

Although many people believe that sleep terrors are just especially bad nightmares, that isn't so. Sleep terrors are completely distinct from nightmares. They happen in different sleep stages and are experienced differently.

Sleep terrors (or night terrors) occur during slow-wave sleep (NREM stages 3 and 4) mostly in the first half of the night, whereas nightmares occur during REM sleep mostly in the second half of the night. If you wake up from a nightmare, you'll probably remember it vividly—or at least parts of it. But people generally remember nothing about sleep terrors. If they remember anything, it's typically just a vague sense of fright, or a hazy image of something chasing them or something frightening in the room.

Bobbi's Story

My son had sleep terrors for several years. They started when he was about 18 months old, not long after he had major surgery—I've always wondered if there was a connection between the two. He would get up, run around the house, scream, and cry. It was like he was having a horrible nightmare, only amplified by 100, because it was as if he were actually trying to run away from the "monster" or whatever he was imagining.

My mother told me that my brothers used to sleepwalk when they were young. They probably had sleep terrors, too, but no one called it that back then. We went to my son's pediatrician and a sleep specialist, and they told us it was night terrors. Well, it was a relief to have a name, but the bad news was that there wasn't much they wanted to do about it. They told us he would grow out of it in a year or two.

We had to make sure the doors and the top of the stairs were secure, for fear that he would fall. The pediatrician told us to lock his bedroom door, but we couldn't do that—we kept thinking the poor kid would just get frantic and bang his head into the door trying to escape!

And of course we felt really powerless because there wasn't much we could do to help him. We followed the old adage about how you should never wake up a sleepwalker, but on the few occasions when he did wake up, he became very disoriented and scared. It was hard sometimes not to just shake him—it's tempting because his eyes were open and he looked awake, so sometimes I wanted to just grab him and say, "Yoo-hoo! Snap out of it!"

The sleep terrors didn't seem to bother my son much while he was awake. He didn't remember them, so the only effect they had was that he sometimes was sleepy in the mornings because he didn't feel like he got a good night's sleep. So the disorder actually affected the rest of us more than it affected him. At the time the sleep terrors started, my second son had just been born, so I wasn't getting any sleep anyway!

This went on for several years—until he was probably around seven or eight, although we'd heard that sleep terrors usually go away by an earlier age. It was a big relief when he finally grew out of it.

CHARACTERISTICS OF SLEEP TERRORS

Sleep terrors are fairly common in children ages 3 to 12, but may start in adulthood or continue all through life. In fact, up until the age of 65 years, the prevalence of sleep terrors throughout adulthood is more than 2 percent in the general population, and then falls to 1 percent. They fall into the same category as confusional arousals and sleepwalking: disorders of arousal usually emerging abruptly from slow-wave sleep.

During childhood, sleep terrors are more common among boys, but during adulthood, they're experienced by both sexes about equally.

Certain autonomic-nervous-system responses are present during sleep terrors: dilated pupils, sweating, increased heart rate (*tachycardia*), and fast breathing (*tachypnea*).

The length of sleep-terror episodes varies. Most instances are just a few minutes long, but sometimes they last half an hour or more. A patient of mine (a woman in her thirties) mentioned that her sleep terrors throughout her life were so intensely horrific that she was convinced her loud and prolonged screaming—which on several occasions caused vocal-cord damage—actually scared away the ghost that was haunting her during sleep and instigating her sleep terror. She emphasized that she was not purposefully scaring away the ghost (since she had no self-awareness at the time), but rather her sleep terror screaming scared away the ghost.

There's normally a theme to the behaviors: fight or flight. That is, those with sleep terrors have an intense motivation to escape or an impulse to "fight back" against a perceived attack. They may punch through walls, jump through windows, forcefully drag someone out of bed (trying to save that person from this imaginary attack), hurt someone nearby because the innocent (usually sleeping) bystander is mistaken for an attacker, pick up weapons, check under beds and in closets, create barricades, smash lamps and other nearby appliances, throw things at invisible bogeymen . . .

The behavior may be in complete contradiction to wakeful behaviors. A quiet, nonconfrontational person may become violent in sleep. A strong, confident person may hide and cry like a baby.

Sleep Terrors and Psychopathology

It's important to note that psychopathology rarely plays a causative role in most sleepwalking and sleep-terrors cases.[1] That is, there is usually no underlying psychiatric affliction causing the disorder. It's just a sudden brain misfire leading to frightened and hyperaroused behavior during sleep. In children, this is probably due to a delay in maturation of their central nervous system, with instability of slow-wave sleep that is also found in adults with sleep terrors and sleepwalking.

For example, in a hypnosis-treatment outcome study that our center reported in 1991, only one patient out of 27 had a current psychiatric diagnosis (panic disorder). Several of the patients had some kind of psychiatric disorder in the past—generally depression, drug abuse, or anxiety disorders—but that didn't coincide with the sleepwalking and sleep-terror onset, and the sleep disorder continued even after the psychiatric problem had ended.

In 1989 our center reported on 54 patients with injurious sleep terrors and sleepwalking.[2] Thirty-five percent of those patients had an active psychiatric disorder and 13 percent had a past psychiatric disorder—most often depression. But as we pointed out in the report, the psychiatric disorders didn't coincide with the onset or progression of sleep terrors; successful treatment of the psychiatric disorders generally did not clear up sleep terrors; bedtime clonazepam therapy promptly controlled sleep terrors and sleepwalking in more than 80 percent of the patients; and nearly all of the patients maintained a high level of psychosocial functioning.

In other words, while there may be an increased history of psychiatric disorders among people with sleep terrors and sleepwalking, the psychiatric disorder is generally *not* to blame, and it is seemingly unrelated to the onset or course of sleep symptoms. When there is a psychological disorder present, getting that disorder under control (through therapy and/or medications) will likely have no effect on the sleep disorder.

Nonetheless, we believe that sleepwalking and sleep terrors are underreported in the adult population because people don't believe the conditions are just sleep disorders. Many of my patients waited years before coming in for an evaluation, and often the explanation is that they were afraid to be told that they're going crazy or having a mental breakdown.

It is true that stress can trigger sleep terror (and sleepwalking) episodes, but stress doesn't *cause* these disorders. That is, if you take a person who is not prone to disorders of arousal and put them under tremendous stress, they still won't have sleep terrors. But if you're already (genetically) predisposed to sleep terrors, stress can bring them out.

Such was the case with David, a patient of mine who had no sleep complaints until he injured his back at work and the assistant administrator began a very obvious campaign to get him fired. Soon thereafter, he began having disturbing, simple dreams about people with blank eyes who were attacking him in some way—following him around, hiding behind stairs with a knife, pointing a gun at him. He'd often dive off the edge of his bed and smash into the wall. Once, his upstairs neighbors heard the screaming and came to check up on his wife. They believed it was her screaming because it was such a high-pitched noise—a sound he says he ordinarily can't make.

But it's the opposite story with another of my patients, a man whose sleep terrors and sleepwalking began when he was in his early sixties and got worse after he retired. By all accounts, he was enjoying his retirement and his stress level had decreased—but his sleep-terror episodes only became more dramatic and frequent. It's unusual for sleep terrors to start this late in a person's life, but not unheard of.

Differences Between Sleep Terrors and Nightmares

As I mentioned previously, sleep terrors usually happen during the first hour or two of sleep during slow-wave sleep, whereas nightmares tend to

happen toward the end of sleep during REM sleep. People with sleep terrors show frequent interruptions of slow-wave sleep ("disordered arousals")
even when they're not having sleep-terror episodes. They seem to have a
lot of difficulty sustaining slow-wave sleep, which is not true of those with
nightmares.

Nightmares are much more common than sleep terrors. Sleep terrors are
not caused by complex dreams; they seem more like instantaneous blips
during sleep that may be associated with simple albeit terrifying imagery
or instantaneous, poorly developed "plots" that then trigger screaming
and other frenzied behavior.

So, while a person may have a long, intricately plotted nightmare leading
up to a fear-associated awakening (and rarely may cry or yell upon awakening), it takes almost no time and minimal stimuli to set off a sleep terror. For
example, in sleep labs, a soft beep from the monitor or another patient coughing in the next room can sometimes set off sleep-terror events in children or
adults who are prone to them. At home, a distant car passing by or the sound
of someone in the house closing a door could quickly induce a disordered
arousal from deep delta slumber. David's wife notes that she tries to be very
quiet when she comes to bed after he's asleep, but he often begins screaming
the second he hears the door open. In other words, there is no prolonged
lead-up to the event, just a hair-trigger response to a small stimulus. This
happens only when the stimulus is applied during a vulnerable sleep stage.

A person waking from a nightmare is feeling upset about the bad dream,
but will have no problem being oriented and recognizing people around
him or her, and will be able to respond appropriately to questions and to
touch. Usually, the person can explain the nightmare in detail. A person in
the midst of a sleep terror is actually still asleep (or, more accurately, in an
altered state of consciousness, stuck in the transition between slow-wave
sleep and wakefulness), can't explain what happened, and doesn't easily recognize familiar people and surroundings—even though the person may call
out for Mom, Dad, or another loved one who is in the room. After fully
awakening, the person probably won't remember the event at all, but may
remember the vague sense that he or she got up during the night, or that
something scary happened. Sometimes, they remember brief imagery.
Rarely do they remember full dreams.

In addition, nightmares are more closely linked to psychological trauma.
Predictably, after traumatic events (which may include surgery, death of a

loved one, abuse, injury, divorce, or many other stressors), both children and adults may experience nightmares during the days or weeks following the event. The nightmares may be about the event itself, or just the emotion of the event, or include some detail of the event in a different scenario. Sleep terrors are not as predictable and do not seem to serve this function, though they can be triggered by stress in those who are genetically predisposed.

I should also point out that nightmares, in fact, more often occur spontaneously, without trauma, in children and adults who do not have anxiety disorder or depression.

Sleep Terrors and Sleepwalking

Most adults who have sleep terrors also sleepwalk. If a person simply sits up in bed and screams, that's a sleep terror—but if that person jumps out of bed and runs around the room, out the door, or through a window, that's a sleep terror complicated by sleepwalking.

Clearly, the latter is more dangerous. There is a "fight or flight" response present with sleep terrors, leading most people to try madly to escape from whatever threat is perceived. It's a frantic effort to get away, *now,* by whatever means necessary—and in the sleep state, that's not usually a logical means. People with sleep terrors have indeed jumped through closed windows, crashed through furniture, overturned beds, and knocked down whatever's in their path. Sometimes they have traumatized their vocal cords from the loud screaming.

One of my patients, who told his story in my documentary *Sleep Runners,*[3] had sleep terrors with sleepwalking for as long as he could remember. One night in his youth, he believed a gang of kids was chasing him and trying to steal his money, so he escaped through his bedroom window and walked onto the roof. He slid all the way down a pine tree and ran down the block in his underwear, where neighbors who were sitting out on their porch spotted him and tried to help. As he regained alertness, he looked down and realized he was covered in blood.

When he got older, he turned to drugs and alcohol as a way of trying to get a decent night's sleep. Nothing worked. He continued having sleep terrors, which disrupted his life greatly. He choked his girlfriend and attacked his sister during episodes when he thought he was fighting an intruder. Finally, he says, something clicked and he was ready to make a change. He quit drinking, and he had been sober for about six months when he came to see me. He was then carefully evaluated at our sleep lab, and the results confirmed my clinical suspicion of sleep terrors. I prescribed clonazepam, and the difference was remarkable: He has not had one sleep-terror episode since then.

It is important to note that this young man's sleep terrors persisted unabated for six months after he "sobered up." In the documentary, he comments that "a mental symptom is not the same thing as a mental illness." This is the most important point to emphasize—the symptoms of sleep terrors can appear to be psychological, and yet they are a manifestation of an intensely altered brain state and not of a deep-seated mental problem. He also talks about the "fine line between sanity and insanity." Living with the daily aftermath of frequent, out-of-control sleep terrors made him question his sanity. Medication restored not only calm sleep but also his sense of psychological well-being during the daytime. It was a rapid turnaround.

Another of my patients described the way his sleep terrors morphed through the years. When he was single and living alone, his sleep-terror behavior was about trying to protect himself—barricading himself into a room, running away from danger. When he married and had children, the behaviors were all about protecting them. He'd frantically try to protect his wife from the imagined intruders, then run for the kids.

Annie's Story

My brother, Luis, has always been a sleepwalker with sleep terrors, and sometimes it's really funny, but sometimes it scares my family and me so much we'd rather keep our distance.

One night, we were staying at a hotel (which used to be a hacienda—really spooky) and we all went to sleep. Then Luis got up and walked all the way to the bathroom. I didn't notice until he screamed. My mom had followed him to the bathroom, and I saw him pointing at an empty corner.

"Mom! The cat! Make it leave, Mom! Look at the cat!" he kept on yelling. The corner was completely empty, yet my brother stared at it, pale and shaking with his eyes wide open.

Mom managed to get him out of the bathroom, and tucked him back into bed. But not even twenty minutes had passed when he sat up in his bed and covered his ears.

"The music! I can't stand the music! Turn off the music, please! The music!" he yelled over and over again, covering his ears and shrinking in bed.

Mom, Dad, and I exchanged confused glances. The room was completely silent; not even a faint noise could be heard. It took us the best part of ten minutes to calm him down again. Needless to say, none of us were able to sleep for the rest of the night.

RISK FACTORS FOR SLEEP TERRORS

In one study of 84 children with sleep terrors (and sleepwalking) from the Stanford Sleep Center, 51 (61 percent) of them were diagnosed with another sleep disorder: 49 with sleep-disordered breathing, and two with restless legs syndrome (RLS).[4] Of those 49 children with sleep-disordered breathing, 42 were treated surgically, which cured the sleep-disordered breathing in all of them, and in every case resolved the sleep terrors (and sleepwalking). Both children with RLS were treated with pramipexole, a dopamine-enhancing medication, and again, parents reported that the sleep terrors stopped along with the presumed precipitating condition (RLS).

So in this study we have children with a genetic predisposition to a disorder of arousal (sleep terrors), but it doesn't come out to bother them until it is unmasked by another sleep disorder that promotes disordered arousals (sleep-disordered breathing or RLS). The interplay between predisposing factors (often genetic) and precipitating factors is a major theme in sleep medicine. Usually, the affected person can have substantial control over the precipitating factors, but not over the predisposing factors (we can't change our genes).

Sleep terrors often run in families.[5] The doctors in the Stanford study suggested that it may be not only that sleep terrors themselves are genetically based but that other sleep disorders (such as OSA and RLS) are also genetically based. They found that 24 of the 49 children with sleep-disordered breathing and sleep terrors had a family history of sleep-disordered breathing.

TRIGGERS OF
SLEEP TERRORS

Sleep terrors can be triggered by stress, sleep deprivation, alcohol use or misuse, and all the other risk factors noted with sleepwalking. Parents often note that a child is more prone to sleep terrors on nights when the child is overtired or goes to bed later than usual.

It's also possible that a full bladder may trigger sleep terrors in children. One parent I worked with brought her child to the bathroom just before bedtime, which stopped the sleep terrors.

In women, we have observed that sleep terrors can occur in conjunction with their menstrual periods.[6] One patient we saw was a 17-year-old young woman who had normal sleep until she was 11 years old (one year after she began having menstrual periods). About four days before menses, every month, she would have an episode half an hour to an hour and a half after she fell asleep.

The first night, she'd sleep-talk and shout. The second night, she'd scream loudly. On the third and fourth nights, she'd scream while sleepwalking and running, often knocking over furniture and attempting to escape through a window. She often injured herself.

She had already tried taking diazepam at bedtime, but it caused over-sedation. We prescribed clonazepam, but that, too, caused her to be too sedated in the mornings. We finally opted for teaching her self-hypnosis, which helped her immediately. When we followed up two and a half years later, she reported that she had only mild sleep terrors about three times a year.

In the second case, a forty-six-year-old woman experienced sleep terrors once monthly, always five to six days before her menstrual period. She would sense a presence in the room or feel the walls collapsing around her, then scream and bolt out of bed, often injuring herself. She had no current or prior psychiatric disorder. We prescribed clonazepam and taught her self-hypnosis, which helped, though it didn't completely eliminate the symptoms. It controlled the sleep-related injuries, but she still reported that less extreme sleep-terror episodes recurred every few months.

Roger Federer, the world's top tennis player, was reported to have had a violent episode of sleep terror on October 6, 2006, in Tokyo. After being asleep, he suddenly jumped out of bed and stood up while screaming. He then took off running and hit the corner of the bed, bruising himself. This was the night before his quarterfinal match at the Japan Open. His fiancée attributed this episode to the stress of playing too much tennis (an ongoing factor), whereas Federer believed it was triggered by the "sake bomber" he had with dinner.

COMMON SLEEP-TERROR IMAGERY

- spiders
- snakes
- monsters, ghosts
- being trapped
- being chased or attacked
- someone attacking a spouse or family member
- robber in the house
- ceiling caving in
- fire burning near the bed

SUPERHUMAN POWER[7]

Often, during sleep terrors, people experience amazing feats of strength and speed. One of my patients kicked a television all the way across the room. Another moves dressers and beds that he can't move at all when he's awake. Often, bed partners and family members are afraid to get too close to the person having a sleep-terror episode because of the fact that they aren't sure they can fight off this "superstrong" person if he or she attacks. They often state that the person with the disorder "leaps" or "flies" out of bed with terrific speed.

Another patient told me, "My roommates in college can tell you stories about my getting out of bed in the middle of the night and moving so fast that the covers were still airborne when I was out the door of the bedroom. . . . You know, you hear those stories about old ladies lifting a car off their son? I think I could literally do that [while having a sleep terror]."

That doesn't surprise me, because we see many patients who have advanced neurological disorders such as multiple sclerosis, and even though they're severely limited in movement while awake, even they can sometimes "fly" out of bed with amazing speed and strength in the grips of a sleep terror or other parasomnia.

I'M AWAKE!

The sleep-terror imagery can be so vivid and convincing that sufferers often believe they're awake. This leads to a frustrating experience when someone says, "Wake up," and the other person insists, "I'm awake!" Those who remember their sleep terrors may find the experience difficult to shake off. One of my patients reports that he would often cry long after he was fully conscious, and another would hunt for family members and try to find evidence of the intruders he was convinced he saw.

One man says, "At times, I have argued the point with my dad. I have said, 'No, no, this was real. This wasn't a dream this time. I did see this.' We would sit and argue for 12 minutes on whether something was there or not."

DISORDERS CONFUSED WITH SLEEP TERRORS

It's possible for sleep terrors to be confused with nocturnal seizures. Seizures are usually shorter in duration than night terrors: about 10 seconds to two minutes long. An overnight sleep study is recommended to monitor for seizure activity if it's suspected. For example, if presumed sleep terrors have unusual features, such as occurring multiple times nightly, emerging late in the sleep cycle, or being linked with the exact same sequence of stereotypical behaviors—such as always sitting up, turning the head to the right side followed by lip smacking, grunting, then screaming—then a nocturnal seizure is possible and should be investigated. An electroencephalogram (EEG) may show seizure-like activity. However, it is well-known that seizures and epilepsy are fundamentally clinical diagnoses, and if a scalp EEG does not show seizure activity, that does *not* exclude the presence of seizures, since seizures can originate deep in the brain and not be detected in a scalp EEG.

Cluster headaches (a form of migraines) may be mistaken for sleep terrors: One published study described four children who seemed to have sleep terrors but who were all diagnosed with nocturnal cluster headaches in the end and successfully treated with the anti-inflammatory indomethacin.[8]

Gastroesophageal reflux disorder (GERD), commonly known as acid-reflux disorder, can mimic a disorder of arousal such as sleep terrors by suddenly arousing the affected person from sleep, who then experiences extreme distress with confusion and disorientation. The person may not readily identify the acid-reflux symptoms, because of the confused state that he or she is in. Also, a small amount of acid reflux can trigger a rip-roaring arousal, which then obscures the cause. GERD usually affects people during the day, besides arousing some affected people from their sleep. Therefore, a careful medical history-taking and "review of systems" should be able to identify GERD, with an index of suspicion that the abnormal nocturnal events may be caused by sleep-related GERD. Therapy at bedtime with medication for GERD should resolve this question.

A doctor should also rule out panic attacks that occur during sleep (*nocturnal panic attacks*). Nocturnal panic attacks, at a superficial level, can look a lot like sleep terrors—both involve a sudden, terrified arousal—but they are very different conditions etiologically. The main differences to note are that a person with a nocturnal panic attack will immediately awaken and recognize his or her surroundings, will be responsive, and will instantly be aware of the panic attack and remember the episode the following day. There is no screaming with nocturnal panic attacks. I have a patient named Tammy who has experienced both recurrent sleep terrors *and* nocturnal panic attacks. Her husband is the frequent observer, and here are their comments in distinguishing these two states: During a sleep terror, Tammy "comes to" only after engaging in her screaming, agitated behavior and running around, and as she regains consciousness, she is disoriented but aware of a racing, pounding heart and terrified state—but she has no idea as to why she is terrified. It takes her a long time to calm down in every respect. Her husband notes a wild look, with dilated pupils.

In contrast, during a nocturnal panic attack, in her sleep (and perhaps in a dream state) she has a brief, vague awareness of a reason for feeling scared and panicky (without ever identifying that reason), which quickly progresses to a rapid and full awakening where she is aware of herself and her surroundings. Her heart is racing and pounding, and she immediately knows she's having a panic attack. Her husband does not observe the "wild look" nearly to the extent that he does during a sleep terror, and it also takes her considerably less time to calm down from a nocturnal panic attack.

In a recent dramatic case, a thirty-five-year-old woman named Denise came to me with lifelong sleep terrors and also daytime and nighttime panic attacks. Her psychiatrist had prescribed the SSRI citalopram (Celexa) which rapidly shut down her daytime and nighttime panic attacks—but had no effect on her sleep terrors that recurred three to five nights weekly for the six months before I saw her. Clonazepam therapy at bedtime then promptly eliminated her sleep terrors. So even though the sleep terrors and nocturnal panic attacks had affected her in similar ways for many years by repeatedly disrupting her sleep in a terror-stricken state with a pounding, racing heart, there were major differences that separated these disorders, besides the divergent responses to therapy: with sleep terrors she screamed, had hypnopompic hallucinations (such as a bedside fan being transformed into an axe that was flying towards her), and would be confused and disoriented before fully awakening. With nocturnal panic, she awakened immediately without any hallucinations, confusion, or screaming. And so this case emphasizes the important point that the concurrence of a psychiatric disorder and a parasomnia in a given person should not imply causality, not even when overlapping symptoms are present, and separate therapies are often needed to control each condition.

TREATMENTS

Sleep terrors often require no treatment, particularly in children. Although they may look scary, they are usually harmless and typically do disappear over time, and often by puberty, in children. (If the problem begins, recurs, or lasts into adulthood, it's more likely to be a long-term disorder in the majority of cases.) A parent confronted with a screaming child with sleep terrors could gently hug (truly embrace) the child for however long it takes until the episode subsides. This type of warm, reassuring body contact can help settle the child, and also allows the parent—who desperately wants to do *something* for the child who is in such distress—to engage in soothing the child and not aggravate the situation.

The first line of defense is proper sleep hygiene. It's important for both children and adults to keep to a regular sleep routine, and to get the recommended amount of sleep every night. Adults with sleep terrors should avoid alcohol and caffeine for at least several hours before falling asleep. The focus should also be on identifying any precipitating factors, such as irregular sleep-wake schedule, sleep deprivation, stress, fever, etc., that can be minimized or controlled in order to lessen the frequency and/or severity of the sleep terrors.

If the person sleepwalks during sleep-terror episodes, safety is the primary concern: Keep sharp or breakable objects out of the bedroom, use a door alarm, barricade stairway entrances, lock main doors, and remove anything the person is likely to trip over.

The next step is to be tested for underlying medical conditions and other sleep disorders. Sleep-disordered breathing may be found in conjunction with sleep terrors, and treating the breathing disorder usually stops the sleep terrors. Additionally, certain medical conditions are reported as provoking sleep terrors. Persistent, high-frequency sleep terrors may also turn out to be a manifestation of a nocturnal seizure disorder. I have often shown a dramatic videotape of a five-year-old patient who was brought to our sleep center by her parents on account of multiple (up to five times!) nightly episodes of what seemed to be classic sleep terrors. We captured several episodes of classic sleep terrors in the lab, including one at 8 a.m.— and all the episodes were clearly nocturnal complex partial seizures, with the EEG pattern showing typical epileptiform activity. Her "sleep terrors" were then completely controlled with anticonvulsant medication.

Benzodiazepine medications, such as clonazepam (Klonopin) are often effective in people with severe sleep terrors. These medications are taken at night (but as with sleepwalking, taken 30 to 75 minutes before falling asleep, to allow sufficient time for absorption and distribution to therapeutic sites in the brain) and presumably work by blunting the hair-trigger response to either internal or external stimuli. Ongoing medical supervision and regular clinic visits are a necessary part of this treatment process. Also, certain antidepressants, such as imipramine (Tofranil) may be prescribed in low doses at bedtime, particularly in children. Psychotherapy at times may be recommended as adjunctive therapy, such as for people who are especially prone to having sleep terrors when they sleep alone at night, and they wish to explore the psychological basis for this connection. More commonly, stress-reduction therapy can be effective in cases where stress is found to play a promoting role.

Self-hypnosis and visual-imagery techniques have been proven effective in children[9] and adults as a treatment for both sleep terrors and sleepwalking. Not everyone is a good candidate for hypnosis. In general, people with disorders of arousal respond to hypnosis, whereas people with insomnia usually do not.

There are various methods for achieving a hypnotic state. This is known as *induction*. You've probably seen a few—like the ever-popular television depiction of a hypnotist swinging a pocket watch and saying, "You are getting sleepy . . . sleeeepy!" But you don't need a pocket watch to hypnotize yourself.

To induce hypnosis, try rolling your eyes upward while lying down with legs uncrossed, then close your eyes slowly while looking up. With your eyes closed, you can follow a brief relaxation technique to relax the body and quiet the mind. Imagine yourself floating in the air or on water, then picture yourself in a peaceful scene (or your favorite place) where you see yourself in a film sleeping restfully all through the night. Any urge to get up triggers a full awakening rather than a partial arousal.

The self-hypnosis exercise takes about 20 minutes and should be repeated twice a day: once during the day and once before bed. During the day, the technique ends with quick restoration of usual wakefulness, but at bedtime the technique ends with drifting asleep. In one study from our center where 27 sleepwalking and sleep-terror sufferers learned self-hypnosis, seven (26 percent) reported no improvement of symptoms, and the remaining 20 (74 percent) reported "much" or "very much" improvement.

IMPORTANT POINTS ABOUT SLEEP TERRORS

The key points to remember about sleep terrors are that they are typically not dangerous to the person's health (aside from injuries that may be sustained while sleepwalking), children do typically outgrow them, they are not usually related to psychiatric conditions, and they can be treated if severe.

Resist the urge to shake or yell at someone who is in the midst of a sleep-terror episode. This can further agitate the person and lead to a prolonged attack. Speak calming words ("You're safe. Everything is okay. You can go back to bed now.") to simply help the person avoid injury. Remember that most people with sleep terrors have no recollection of them, so they're generally more upsetting to parents and bed partners than they are to the person with the disorder!

Chapter 11

SLEEP PARALYSIS

*Dreaded Visits from
the Old Hag*

Imagine falling asleep with fully mobile limbs, only to awaken inexplicably paralyzed. However you try, you're quite unable to move—not the wiggle of a toe, nor the stretch of an arm, nor the quiver of a lip.

And you have good cause for lip quivering because there's something else in the room with you—something sinister, something otherworldly. Whatever it is, it's definitely not good. It can loom over you and then pounce on your chest, crushing the breath from you.

Dread splashes over you, flushing you tip to toe. You try hard to stir your stunned nerves to action, to recall your incapacitated arms and legs, but it's like shaking off the coils of a snake. It's impossible, and that thing on your chest is getting closer and closer to cutting off your air supply . . .

Welcome to the exquisite terror of sleep paralysis!

WHAT'S SLEEP PARALYSIS?[1]

Sleep paralysis is a period of head-to-toe immobility, except for the eyes, lasting for as short as 30 seconds or for longer than five minutes. Slumberers struck by SP find their breathing more labored than under normal conditions. They experience fear, this feeling enhanced by their inability to command their limbs, their difficulty breathing, and the sense that something or someone is in the room—often mythically characterized as old hags, incubi, succubi, or other such supernatural creatures. The creature is usually interpreted as something scary that may want to suffocate the person or suck out their breath.

These hallucinations can occur during the transition from wakefulness to sleep (called *hypnagogic hallucinations*), or during the transition from sleep to wakefulness (known as *hypnopompic hallucinations*).

People experiencing SP may feel tingling or shaking sensations, or brilliant explosions of light or feeling inside their heads. Some people may see tunnels (similar to a white-light experience), hear buzzing or ringing sounds in their ears, feel like they're floating, have a sensation like an out-of-body experience, or (rarely) fall into a blissful state.

Sleep paralysis is probably the most fantastic and lore-ridden of all parasomnias. Narcoleptics, sleep-talkers, and sleepwalkers often find themselves caricatured on sitcoms and movies, but sleep paralysis is something special and mysterious, a straddling of the threshold between inner fantasy and outer reality, while the sleeper himself is sometimes straddled by a bizarre, hallucinated visitor. Many studies have drawn parallels between SP and extraordinary phenomena such as demon possession, alien abduction, shamanistic trances, and white-light and out-of-body experiences.

What's fascinating about sleep paralysis is how shrouded it still is in supernatural phenomena. In many cultures, sleep paralysis is still referred to by a mythical title rather than its mundane scientific classification. But when sleepers' dream lives seem to crawl, half-formed, out of their own imaginations and into their bedrooms, it must be awfully tempting to believe something mystical or miraculous is happening, and it's not just due to daily stresses or poor sleep posture or some other idiosyncratic explanation.

Sometimes, it's hard to convince people with sleep paralysis that their houses aren't "haunted." Some are certain that they've angered a ghost who now wants to suck their breath out while they sleep. In other cases, entire books have been written about alien abductions that most sleep doctors would consider episodes of sleep paralysis. The experience of sleep paralysis is so hyper-real and yet so out of the ordinary that affected people sometimes feel compelled to write extensively about what they believe is a supernaturnal experience that can repeat itself multiple times over a period of weeks or months.

Sleep paralysis was first described by science in 1876, by American neurologist Silas Weir Mitchell, who captured the condition with the following description:

> The subject awakes to consciousness of his environment but is incapable of moving a muscle; lying to all appearance still asleep. He is really engaged in a struggle for movement, fraught with acute mental distress; could he but manage to stir, the spell would vanish instantly."[2]

Since that inaugural paragraph was published, numerous studies have been conducted on SP and its many incarnations in modern society. From the data generated by these studies, we have learned that SP is experienced by people across continents and from all walks of life. SP strikes the young and the not so young, male and female, people of all races and cultures. And even if the words and explanations we use are different, the experience of sleep paralysis is remarkably similar among all people.

Sleep paralysis can be a symptom of narcolepsy, or it can appear on its own, called *recurrent isolated sleep paralysis*. This may happen a few times in a lifetime, or it can happen almost nightly. There is enormous variability in the expression of sleep paralysis.

Isolated sleep paralysis is a fairly common condition. It's generally accepted that about 25–30 percent of the population will experience an episode of sleep paralysis at some point.

WHAT SP FEELS LIKE

As I mentioned, an SP attack occurs either on the verge of waking up or just as the subject is falling asleep. Not all SP attacks are "full-blown"—that is, they don't always include all possible stages, but the experiences associated with SP are as follows:

- Either during the hazy state just before sleep or on arousal, the person's body feels heavier and heavier. This may be accompanied by a tingling, buzzing, or vibrating sensation.
- Now there is a total inability to move or speak. The person is awake and aware of his or her paralysis. He or she may try to call out for

help, but nothing happens. It's possible but unlikely that the person can get out a whisper or make a quiet noise. Fear sets in.

- It becomes hard to breathe, and there is a feeling that something is pressing down on the chest, constricting the lungs. Few people report actually seeing something sitting on their chests, but they often sense it. It may also be described as a choking experience. They may believe they are being punished or are the victims of angry spirits.

- A repetitive sound is heard in the ears—hissing, ringing, popping, etc.—growing louder. "I thought my eardrums would burst," one sufferer described.

- The presence of another being—often regarded as supernatural— is felt, heard, or seen by about half of all people experiencing SP. This intruder may be accompanied by flashes of light, or music, or voices. In most cases, it's believed that this intruder is evil—and just out of the range of view. "If I could just turn my head two inches, I know I'd see it!" the sufferer swears.

- Other sorts of visual, tactile, or auditory hallucinations may occur. In extreme cases, people may feel that they've exited their bodies and can travel around the house, around town, or anywhere else. For most people, this is frightening, but it can also be a portal into lucid dreaming—a state where you have control over your dreams and be aware that you're dreaming—and that can be a pleasant and even exciting experience.

- The person tries to move a body part or scream to break free from the paralysis. It may work, or the feeling may eventually go away on its own.

- Upon coming out of the attack, people generally gasp for air, sweat, and feel a racing heart. It can take some time to recover from the event afterward, particularly for the first few attacks when the person has no idea what's happening. Once the attack is broken, the same sensations may be freshly experienced each time the subject tries to fall asleep.

The most common form of sleep paralysis is its simplest: just an inability to move or speak, and perhaps a vague sense of doom. Those who experience full-blown attacks that include hallucinations do not necessarily experience those attacks every time. Some sleep-paralysis instances may be short and uncomplicated; others may be long and fraught with frightening imagery, sounds, and sensations.

Paralysis during REM sleep is a perfectly natural function of the human body, a special mechanism designed to keep the dreamer from acting out his or her dreams in real life. However, in an SP attack, most or part of the waking consciousness is maintained after the paralysis experienced during REM sleep has kicked in.

My SP Story

I experienced recurrent isolated sleep paralysis for about two years, about one to three times per month, when I was seven and eight years old. Out of the blue, without any stress or medical problems, I started having states of sleep paralysis early in the morning, not long before I was supposed to get up for school. Each time was always the same, with the first time being especially strange and frightening because I had no idea what was going on even though I could feel what was going on—a large mountain was sitting entirely on top of me, and I had no chance of moving any part of my body. I could breathe, but it was not easy; otherwise I felt fine and rested from a good night of sleep. I just couldn't move and get out of bed and start my day.

It was a strong feeling, without visualization, that there was a mountain on top of me, a regular large mountain with some trees on it. I accepted without question that this is why I had paralysis. That was the way it was, in the same way that dreaming has a major noncritical component: We accept the dream as de facto reality. The body feelings and overall feeling were so strong that all of my senses understood that it was a mountain on top of me, even though those senses didn't themselves process the mountain in their own modalities. Although I had no visual hallucinations at this point, I did have tactile hallucinations with my SP—a massive crushing sensation, without injury or any feeling of physical trauma.

And then, maybe six months or so later, in a gradually evolving fashion, I could start seeing the very high and big mountain (and eventually saw some trees and shrubs on the steep slope) during most episodes of SP together with my usual crushing tactile hallucinations. I could see the green and brown and other colors on the mountain, but no animals or birds. I was at the bottom of the mountain (of course) and always looked high up. The sky was always blue with some passing clouds. So then I had two forms of hypnopompic hallucinations with my SP: tactile and visual. I had enough self-awareness in my hypnopompic mode to say to myself, "So *that* is the mountain that has been crushing your chest and keeping you from even twitching a muscle."

My self-awareness was not that from the quasi-wakeful mode, realizing that I was hallucinating and had been asleep and was soon to wake up—instead, my self-awareness was that of saying within my dream-hallucinatory mental state, with full concrete conviction, "That is the actual mountain that has been crushing and paralyzing you." (In retrospect, it almost pushed me to the point of trying to name that mountain—it was becoming that familiar.) Things started making more sense to me because I could see the big thing that was crushing me as I was about to wake up.

I was always lying on my back (meaning I had sleep posture–specific SP), I never screamed, and I never told my parents or asked any questions. I just knew after the first episode that it was "one of those things" that was not really serious, and it didn't bother me that much. And finally it all stopped, all of those weird morning things stopped. I hardly thought about it anymore.

Much later in my life, I had almost daily auditory (and, rarely, hazy visual) hallucinations for several months upon waking up in the morning, but without any SP whatsoever. This happened after my father died in 1991. I could hear his voice waking me up in the same gentle and firm manner (tone, volume) as it did in my childhood to get me up and going and ready for school. Some mornings, I could see his face in a distant way. His voice, though, sounded so real, just like in the early years of my life. This phenomenon of hallucinating a lost loved one has been fairly well described in psychiatric publications.

"For what seemed ages piled on ages, I lay there, frozen with the most awful fears, not daring to drag away my hand; yet ever thinking that if I could but stir it one single inch, the horrid spell would be broken. I knew not how this consciousness at last glided away from me; but waking in the morning, I shudderingly remembered it all, and for days and weeks and months afterwards I lost myself in confounding attempts to explain the mystery."

—From *Moby-Dick* by Herman Melville

Josh's Story

I've always felt extremely claustrophobic inside my body. As a kid, I was anxious about that moment when my conscious mind canceled out, my body shut down, and my dream world flared to life. I'd lie awake for hours like a condemned man, feeling just as helpless and vulnerable. My early fears of being locked in my body with the key just beyond my grasp took on a horrific manifestation.

It started when I was eight, shortly after I realized my own mortality. I was suffering through a bout of highly symbolic nightmares of the end of the world, when, at the peak of my terror, just as the cleansing fires were about to wash over me, I would jerk awake, ready to bolt, only to find my deepest fears realized: my limbs inflexible as petrified wood, every nerve and muscle fiber AWOL.

It would feel as though the malicious hordes were mounting the stairs outside my bedroom. I could picture their jackboots thumping and their shoulders clanking with the tools of their terrible trade. They were at the door, snorting like razor beasts, sanguine smoke curling from under the door, like the portal to Hell in a production of *Don Giovanni*. Then, just as the lock snapped . . . I had my body back, heart pounding, lungs gratefully slurping in air, as though I'd been trapped underwater.

This was my first sleep-paralysis experience. More followed. When anxieties press into my temples and pour into my gut, sleep-paralysis episodes follow, sometimes 20 episodes a night.

It's an ache I have to live with, a daily threat. On particularly bad nights, I forgo sleep entirely.

WHAT CAUSES SP?

Before this most modern age of science, reason, and logic, many cultures turned to fantastic reasons to explain sleep paralysis. Even today, many people turn to the supernatural when asked to define SP. Truth is, we still don't know why it happens. The best we can do is to describe the physical and psychological events that surround it, but it's still difficult to explain why large numbers of people all around the world with sleep paralysis share the same types of hallucinations.

Sleep paralysis is most likely caused by inappropriate intrusions of the atonia (lack of muscular tone) of REM sleep into wake-sleep and sleep-wake transitional states. It's believed that the atonia of REM sleep either sets in too early or hangs around too long, along with the associated dream state (predisposing to hallucinations)—despite wakeful consciousness. This characteristic often occurs with people who were recently sleep-deprived, who are feverish, or who have been aroused from deep sleep. Therefore, in the same way that reality bends and twists in our dreams, these hallucinations may be in some way based on reality but twisted. For example, an object in the room, or a shadow, may take on new shapes and "come alive." The sound of the wind in the distance may become voices in conversation.

The degree of hallucinatory experience varies from person to person. Some are absolutely convinced that what they see and experience during sleep paralysis is real; even after coming out of the experience, they remain convinced that someone or something really was in the room, or that everything happening in their minds was happening objectively. Others know that even though it all *feels* very real, there's something wrong with the picture, and it can't be happening the way it seems. These experiences usually aren't foggy and gauzy, like some dreams, and they don't feel like the confused visions you might get when you have a high fever or have taken too much cold medicine. They feel plainly real.

If you read accounts from people who are sure they've been visited by ghosts or aliens, many of them are textbook cases of sleep paralysis, yet this doesn't seem to lessen their belief that otherworldly creatures are stalking them. Some people develop strong belief systems about why these creatures must be targeting them, and complicated rituals to ward off the creatures—

such as performing spells to drive spirits away and burning sage to "purify" their homes.

In a fascinating study in Japan, sleep researchers brought 16 subjects into a lab and purposely tried to elicit SP by interrupting their sleep at strategic times.[3] The researchers wanted to induce sleep-onset REM periods, meaning that the subjects would enter REM sleep immediately, or almost immediately, after falling asleep, going against the body's natural tendency to enter sleep through stage 1 NREM sleep. They succeeded in producing isolated sleep paralysis six times.

In all cases, the subjects reported feeling an inability to move and an awareness that they were lying in the laboratory, and all but one had auditory or visual hallucinations and unpleasant feelings. All but one of these sleep-paralysis events happened during a sleep-onset REM period, which led the researchers to conclude a strong correlation between sleep paralysis and entering sleep in the REM stage.

This reinforces the idea that SP is closely connected with "faulty" REM sleep mechanisms. We're most likely to experience these periods when we're sleep-deprived, napping, sick, feeling stressed, or have an irregular sleep schedule. Therefore, all these times are ripe for SP attacks.

It's possible that the bodily symptoms provoke fear, which brings on hallucinations because we *want* there to be an explanation—the mind may create sensations of otherworldly presences to provide a reason for the bizarre paralysis.

In a 2004 report, Richard J. McNally and his team of researchers showed that a group of ten people who believe they've experienced alien abductions showed signs of post-traumatic stress disorder.[4] The researchers taped the subjects' accounts of their experiences, then played the tapes back for them. When the subjects listened, their pulse rates went up, their palms became sweaty . . . they showed the same symptoms that a war veteran might show when remembering trauma. The researchers concluded, "Belief that one has been traumatized may generate emotional responses similar to those provoked by recollection of trauma."

Dr. Michael Shermer, founding publisher of *Skeptic* magazine, analyzed this study in *Scientific American*. For a "professional skeptic," he has an interesting background: He had his own alien-abduction experience. In 1983, he believed that alien beings forced him onto their spacecraft for about 90 minutes.

"My abduction experience was triggered by sleep deprivation and physical exhaustion," he wrote in the February 2005 issue. "I had just ridden a bicycle 83 straight hours and 1,259 miles in the opening days of the 3,100-mile nonstop transcontinental Race Across America. I was sleepily weaving down the road when my support motor home flashed its high beams and pulled alongside, and my crew entreated me to take a sleep break. At that moment a distant memory of the 1960s television series *The Invaders* was inculcated into my waking dream. In the series, alien beings were taking over the earth by replicating actual people but, inexplicably, retained a stiff little finger. Suddenly the members of my support team were transmogrified into aliens. I stared intensely at their fingers and grilled them on both technical and personal matters. After my 90-minute sleep break, the experience represented nothing more than a bizarre hallucination, which I recounted to the television crew from ABC's *Wide World of Sports* that was filming the race. But at the time the experience was real, and that's the point. The human capacity for self-delusion is boundless, and the effects of belief are overpowering."

I'm not saying that it's impossible that aliens have visited Earth, or that ghosts have invaded people's bedrooms, but what I am saying is that I hope more people will look at sleep paralysis first, not last, in the list of likely explanations for nocturnal visitations.

MYTHS AND LEGENDS ACROSS CULTURES

Some ethnic groups report that as much as 62 percent of their population experience this condition. Although it's not certain why there are vast differences among ethnic groups, it may be that people report having the experiences more often when they know what it is—if you ask Americans if they know what sleep paralysis is, a significant number will have no idea what you're talking about. But if you ask Korean young adults if they know what *kawinullium* is, the vast majority of them have heard the term. It means, roughly translated, "ghost entering the body and squeezing or pressing," and that's their mythical term for sleep paralysis.

In a new study involving men in the Korean army, 34 percent of subjects said they had experienced SP at least once. So it's possible that they're remembering these experiences better because they've grown up with folklore about it and recognize it when it happens, whereas people who've never heard of such a thing just think of it as a freaky occurrence and forget about it after a while.

Sleep paralysis is often experienced together with hypnagogic or hypnopompic hallucination, and people of different cultures have different names and explanations for it. In Germany, sleep hallucinations are known as *Hexendrucken,* or witches passing through the sleeper's bedroom. In Mexico, they are *pesadillas,* or nightmares. In China, they are called "ghost oppression." People in Thailand know sleep paralysis as *phi um,* where the sleeper is consumed by the body of a specter. The Japanese call this phenomenon *Kanashibari.*[5, 6] In St. Lucia, one might hear sleep paralysis referred to as a *kokma,* an assault by the ghost of a baby, straddling victims' chests and choking them. In the United States, in African-American culture, sleep paralysis is defined as a witch "riding" her victims while seated on their chests. A psychiatrist working with Cambodian refugees found that most of them attributed sleep paralysis to the angry spirits of people who had died unjustly and were now taking it out on the living. In Japan, nearly half of all male college students and almost two-thirds of all female college students suffering from sleep paralysis believe in a possible link between the paranormal world and their experiences with the condition.

Newfoundland's Old Hag

In 1982, David J. Hufford published *The Terror That Comes in the Night,* a highly regarded book that explores the "Old Hag" phenomenon in the Canadian province of Newfoundland and other cultures.[7] While a teacher at Memorial University, Hufford became intrigued by the responses to a questionnaire distributed to students in 1970, asking them to describe their Old Hag experiences or to interview others who'd had these experiences.

Two students interviewed a 62-year-old woman who detailed a love triangle between three people she knew in 1915. John and Jean were a steady couple, but Robert (a schoolteacher) wanted to date Jean. The woman explained that "about a month after this had been going on, Robert began to be hagged." She says that every night, Robert felt as if someone were pressing against his chest and strangling him. He was always lying on his back, and he couldn't speak when this "intruder" came into the room. He was sure it was John who was hagging him, out of jealousy—that he'd put a curse on Robert to steal his breath and his voice.

The account continues, "The way to call a hag, Robert later learned, was to say the Lord's Prayer backward in the name of the devil. The only way to avoid the hag was by drawing blood or using the word *God* and keeping the light on in the bedroom."

The same method for calling a hag—saying the Lord's Prayer backward— was backed up by an 80-year-old man explaining how his friend hagged a woman who refused to kiss him. The man also explained how to kill a hag: "If you swing at a spirit with a bottle, the spirit who is haggin' you will die."

Hufford concludes that in Newfoundland, an Old Hag is thought of as a supernatural creature (like a vampire or ghost), a human being acting in a supernatural way, or a combination of the two. It may also be regarded as caused by indigestion or circulatory problems.

In some parts of Europe and Great Britain, people to this day collect "hag stones," or "holy stones," which are stones with natural holes running through them. The tale is that hanging a hag stone at the head of your bed or keeping it under your bed will protect you from the evil spirits that come to crush you as you sleep.

WHO GETS SP?

There are several key features that make people likely to experience SP attacks:

1. **Adolescence.** Most subjects experience their first SP attacks when they are teenagers, with the "magic number" frequently cited as age 17. It's rare for a first attack to happen after age 25.
2. **Anxiety.** Certain anxiety disorders, specifically panic disorder and post-traumatic stress disorder, may promote SP attacks. Also, fear of death or a belief in controlling fates or luck may be influencing factors.
3. **Shift-working.** A study of 8,162 people in Japan found that people who work night or evening shifts are much more likely to experience SP than those who work nine to five. The disruption

of sleep timing mechanisms associated with shift work can lead to dysregulation of REM-sleep motor mechanisms during sleep-wake transitional states, resulting in SP for those who have a vulnerability to this phenomenon.

4. **Sleeping Position.** People who sleep in a supine position—that is, lying flat on the back with the face directed toward the ceiling—may be more likely to experience an SP episode.

SP WITH NARCOLEPSY

While SP can, and often does, occur as an isolated phenomenon, independent of some larger pathology, SP is one of the four symptoms of the "classic narcoleptic tetrad."

Narcolepsy, a disorder full of dissociated sleep-wake states, is the most notorious instigator of recurrent episodes of sleep paralysis. Thus, SP and hypnagogic/hypnopompic hallucinations often affect a person who is afflicted with the two core symptoms of narcolepsy: cataplexy and daytime sleepiness. Although SP can occur as a symptom of narcolepsy, it is not present in all people who suffer from narcolepsy.

SP PREVENTION AND TREATMENT

Since SP occurs as both a symptom of narcolepsy and as an independent condition, it's difficult to come up with an all-purpose medical approach to cure or prevent it. As with all illnesses, the approach to take depends on the individuals and their needs, especially since SP has a large psychological and cultural component.

There is no medical need to prevent SP; the main reason to treat it is simply to ease the concerns and fears of those who dread the feeling. Also, some people with recurrent SP, or some other parasomnia, may eventually develop *anticipatory anxiety* about going to sleep, which in its most severe form can result in frank "sleep phobia." I have seen a number of patients over the years with these secondary problems, and so I know how much better it is to prevent these secondary problems from happening rather than to control them after they have developed.

Many individuals suffering from SP fail to talk to a doctor about it, perhaps worrying they might be regarded as insane. It's true that in the past, recurrent SP was prone to being misdiagnosed as a mental illness. However, we now know that SP is not an indicator of mental illness. It may be a sign of narcolepsy, however, so a consultation with a sleep doctor may be a smart idea. The doctor should provide the patient with as much information about the condition as possible, as this might help ease the patient's anxieties, which could very well be a strong provoking factor in SP attacks.

People suffering from recurrent isolated SP, or SP associated with narcolepsy, may wish to try some of the following suggestions, or they may adopt original approaches that suit their needs. (Keep in mind that there are effective medications available, which will be discussed in the next section.)

TECHNIQUES TO STOP SP

You may wish to try the following methods to prevent or break free from an SP attack:

- **Alter your sleeping position.** As I've mentioned, people who sleep supine (on their backs) are far more likely—five times more likely, according to Allan Cheyne, a top expert on sleep paralysis and psychologist at the University of Waterloo in Canada—to experience SP attacks. Some people may experience SP attacks no matter how they lie in bed, whereas other people may have an SP attack only when lying in a certain position, be it on their left side, on their right side, prone (on their stomachs), or supine, so a change to one of the other sleep positions will most likely do some good. Of course, for just about everybody, there is some tossing and turning involved throughout the night, so a person may fall asleep lying on the stomach but end up lying on the back. A way to prevent this is to attach tennis balls or similar objects placed in homemade pockets to the back of the pajamas, as I've suggested for sleep-apnea sufferers. When you roll over during sleep, the discomfort should cause you to instinctively roll out of the position.
- **Make small motions.** When they discover themselves in the grip of an SP attack, many people attempt to struggle out of it to no avail. The instinct is to sit up, or at least flail the arms or kick the legs, but this is a monumental task when you're effectively paralyzed. One

possible method of escaping the petrifying womb of an SP attack
is to make small movements instead of large ones. Your bigger
muscles may be frozen, but it may be possible to direct the smaller
ones—the fingers and toes in particular. Often just one small
movement of a finger is enough to break the paralysis. If you can't
move your fingers, try shifting the eyes back and forth.

- **Cough.** If you can't move at all, try coughing or making a noise
  like a groan or a hum.
- **Take medicine.** Medications affecting REM sleep and its motor-
  behavioral components (including muscle tone) can help control SP.
  Specifically, in my experience, the tricyclic antidepressant imipramine, taken at bedtime, often in low doses, can be very effective and
  well-tolerated, and can bring welcomed relief. Sometimes higher
  doses are required. Other tricyclic antidepressants, such as clomipramine, can also be effective, along with certain other classes of
  medications. These medicines have REM-sleep-suppressing properties, so they can help overcome SP by eliminating any stray "bits
  and pieces" of REM sleep (such as muscle paralysis) that intrude
  inappropriately into wakeful states. Also, if you believe that your
  SP began with, or became aggravated by, the start of a new medication, then you should check with your doctor to discuss this
  matter and the question of a different medication.
- **Go back to sleep.** If you're unable to break free from sleep paralysis and it's unnerving you, try just letting yourself fall asleep again.
  Most likely, you'll be just fine next time you awaken.
- **Avoid naps.** You're more likely to experience SP during daytime
  naps, or when you're sleeping in an unfamiliar place. So consider it
  a double whammy if you're napping in an unusual place, like on a
  friend's couch or in the student lounge.
- **Follow your beliefs.** Some people who firmly believe that this is
  a spiritual problem choose to combat it in spiritual ways. Whether
  I believe that or not doesn't matter—the mind is powerful, and if
  you want to try using the tools of your spiritual beliefs to end attacks, you should certainly feel free to try. Some sufferers report
  that praying during an attack, calling their soul home, or professing
  their faith to the otherworldly creature puts an end to the episode.

- **Learn to relax.** Stress is one of the primary culprits behind SP. Not only can stress itself precipitate SP, but when we're stressed, we tend to smoke or to drink too much coffee or alcohol (and at the wrong times); we may stare at a computer monitor late into the night, and eat "comfort foods" like chocolate (which contains caffeine) and sugary snacks, and this wreaks havoc on our sleep system. Exercising restraint or being abstinent from these substances and activities will most likely benefit your sleep life. You may wish to incorporate relaxation techniques such as meditation into your daily life. Guided-imagery recordings meant to promote relaxation are available on CD or tape; a speaker will guide you through imaginary scenarios like a walk along a beach or flight through the clouds. Also, a good way to keep daily stresses from interfering with your sleep is to maintain a regular sleep routine—go to sleep and wake up at the same time every day.

Reassure Yourself

Knowledge may be the best way for the SP sufferer to arm himself or herself against the condition. Knowing that an SP attack isn't an instance of demonic possession or an antagonistic act of witchcraft—that it's simply a physical condition based on a badly timed sleep cycle (specifically, the inappropriate intrusion of the normal paralysis of REM sleep into our sleep-wake transitions)—helps the sufferer realize that there is nothing to be ashamed of, nothing freakish or uncommon going on.

Sleep paralysis is not dangerous, and not a sign of impending insanity or death. It may be scary, but it's temporary. Rather than fighting through the attacks, many people have learned to tolerate and even enjoy the experiences by "floating" through them and seeing them as interesting phenomena rather than something scary.

Lucid Dreaming

A small percentage of those who experience SP actually look forward to the experience because they use it to explore parts of their subconscious minds that are not usually accessible. From the SP state, it's possible to enter a dream state where you're aware that you're dreaming and can even control the content of your dream if you wish.

If you're feeling adventurous, you may choose to fly around the room or out the window, explore worlds of your own creation, or move through walls and mirrors. Sometimes it's possible to control every facet of a dream, and other times you're more of an observer and the dream morphs of its own accord.

There are many good resources about lucid dreaming if this topic interests you. Numerous books are published on the topic, as well as websites such as www.lucidity.com and www.dreamviews.com.

WHAT "NIGHTMARE" REALLY MEANT

What we call a nightmare today is not the same experience as what was originally meant when the word was coined. *Nightmare* originally described sleep paralysis. Mike and Melanie Crowley, editors of the *Take Our Word for It* word-origin webzine, say the word comes from the Old English word *maere,* which meant goblin or incubus. They write, "The word was *nigt-mare* in 1300, and it referred to an evil female spirit afflicting sleepers with a feeling of suffocation." They say it was used the way it is today—to describe a bad dream—by 1829.

SP AND HOW YOU VIEW IT

The first SP attack is usually the scariest. The person has no idea what's happening or whether they'll ever break free—some believe they're going to die (or are already dead). Although further episodes may still be terrifying, there is at least the assurance that they've lived through it before and that the episode ended.

After that point, different personality types react in different ways: Some people worry that they're losing their minds; others feel a tremendous sense of guilt because they think the spirit is punishing them for something they've done; some purposely try to prolong the experiences and see what they can learn; some find it frightening while it's happening but still interesting to think about later; some immediately start calling literary agents and film producers to tell their tales of alien abductions; and some sink into depression or anxiety.

Because people have different outlooks on what's happening and there are different severities of the disorder, not everyone wants it to be treated. But those who dread the experience and have it frequently should not feel ashamed to seek professional help.

Chapter 12

SLEEP-RELATED BINGE EATING

Relentless Loss of Control

Ⅰt was almost like packing a bag lunch for school, Amy Koecheler says. Every night, she or her mom would set out the snacks she'd eat that night—like chips or pudding. But this was no ordinary bedtime snack: These were foods Amy would eat while sleepwalking.

Amy's mom, Shirley, can't remember a single night in her daughter's life when Amy slept through the night without getting up to sleep-eat. She says that even in the womb, Amy was unusually active, constantly kicking and moving. And as soon as she could walk, she began sleepwalking and sleep-eating.

When Amy was about 13 years old, her mom found her asleep on a couch with chocolate chips smeared all over her mouth, in her hair, and smashed into the couch. Although it was 4:00 a.m. at the time, Shirley had to wash Amy's hair—she couldn't send her back to bed with chocolate all over the place! Shirley may have understood better than most, though: She was a lifelong sleepwalker, too, and has multiple awakenings through the night. They both take Amy's disorder with a good dose of humor, but it has caused practical problems and social embarrassment. She first came to see me in 2002, when she was about to go off to college and face life with a roommate. Little did she know that her roommate was going to be a sleepwalker too. (They decided to forgo the bunk beds!) Amy also suffered from lifelong RLS. After our evaluation, I prescribed nightly treatment with a dopamine medication, a benzodiazepine medication, and the anticonvulsant topiramate, which controlled her RLS and her sleep-related eating disorder (SRED).

SRED has received increasing amounts of media attention in recent years. Many of our patients came to see us only after seeing a report on television or in a magazine describing the problem: They had no idea that many others shared this condition and that it was treatable. At least 1 percent of Americans have SRED; another 1 to 2 percent have night-eating syndrome (NES), a disorder of abnormal wakeful eating at night; and still others have nocturnal bulimia nervosa and nocturnal dissociative disorders with "eating personalities"; so we are talking about more than 3 million people and even 4 to 10 million more who may be affected by such conditions.

Donna's Story

I have been a sleep-eater ever since I can remember. I think it runs in my family because my dad would call it a "midnight snack"—I remember sneaking out to the kitchen and joining him for peanut butter on toast and hot cocoa. Then it was chocolate ice cream in the middle of the night, and as I got older, it was anything that was sweet and rich.

I have to travel from my bedroom, which is in the basement, to the kitchen to do it. Sometimes I subconsciously tell myself, *You don't need to do this. You're not hungry,* but nothing stops me. I can't stop myself. There is no willpower. There are times when I'm not conscious of it and times when I'm semiconscious of it, but it's an unstoppable drive.

I know I can't buy cookies or cake—if they're in the house, I *will* get up in the middle of the night and find them. I've literally taken a fork and eaten half a cake; I'll wake up in the morning and find half the cake is gone, and I had subconsciously done it. So I learned to hide food from myself.

I've tried to limit myself, to take just five or six little Hershey's candy bars and line them up on my chair—but I still get up and find more because I know there's more in the house. I'm diabetic, so this is dangerous. Then I figured I'd outsmart this by just eating hard candy . . . but I woke up with hard candy in my mouth, and that scared me. I could have choked to death! Big mistake. So I started keeping food in the garage, but I'll make my way to the garage, even in the middle of the night in the dead of winter, to find candy in the freezer in the garage.

I bought a box of sugar-free chocolates and I went through the whole box in my sleep. I must have taste-tested every one of them. I woke up the next morning, and I had a chocolate taste in my mouth, so I looked in the box. They must have been pretty lousy because I had taken a nibble out of every one, and I put them all back in the box!

It's the most frustrating, defeating feeling to know that you've done it again. It's almost like an addiction. I'm an addict in recovery, so I know what an addiction feels like. I'd wake up with piles of wrappers in bed; chocolate melted on my face, on my sheets and pillowcases, in the pocket of my robe; peanuts in my mouth when I had candy bars with peanuts . . . it's a horrible feeling to wake up and know you couldn't stop yourself.

My 15-year-old daughter is starting to manifest the same habits, and it has been getting worse and worse. My 23-year-old son has been doing it for years. He started out with caramel sauce in bed. When I went to visit family years ago, my nephew asked why I was on the medications I was on, and I told him about my sleep-eating. He said, "Aunt Donna, I woke up with a Snickers bar in bed the other night." My sister and I have talked about it, and she says, "I try to eat a bagel before I go to bed so I don't wake up and eat it in the middle of the night." So it's something I believe runs in my family, but I'm the only one who's ever done anything about it . . . and that's because of a fluke! One of my children had bed-wetting and night terrors, and I went to see a sleep physician about that. While I was there, I mentioned my sleep-eating, and he asked me to come for testing.

I was also recently diagnosed with hypersomnia, which is excessive sleepiness. I've been to the point where I was totally exhausted and I couldn't get up off the couch, so I finally called Dr. Schenck and said, "I can't get up anymore. I'm sleeping all day long!" I've been treated for ten years for the sleep-eating disorder, but now I'm starting to have manifestations of other sleep disorders.

When I was first diagnosed, I took clonazepam, levodopa, and Tylenol with codeine. But when I was in recovery, I wanted to stop the clonazepam and Tylenol #3, and I had some facial twitches from taking the levodopa alone. So he tried me on Topamax along with my nightly trazodone. We started on a low dose and steadily increased it by 25 mg at a time. I've been on it for about a year and a half, and so far, I've had really good success with it. And I've lost about 30 pounds in the last few months.

I know if I don't take my medication, I will be in the kitchen that night. Cookies, candies, pizza, peanut butter . . . whatever is good and rich and fattening, I will find it. If I do take my meds and I'm really stressed out, I may still have a breakthrough. It's something I have to live with, but as long as I follow my treatment and do what Dr. Schenck tells me to do, I can keep it under control.

It's a relief to know that I'm not the only person out there dealing with this. When I first talked to my doctor about it, I thought for sure that he would think I was crazy. You don't discuss those kinds of things. You keep it buried inside. Now that I know what this is and that there are medications that can help, I try to talk about it with my friends. Some will say, "I've been doing that for years!" and I say, "It's really something you should get checked out. You could find a lot of relief from it."

HISTORY OF SRED AND OTHER NOCTURNAL EATING DISORDERS

Night-eating syndrome was first described in medical literature in 1955 (just two years after the discovery of REM sleep) by Dr. Mickey Stunkard and colleagues at the New York Hospital clinic for obesity. In a group of obese patients who had not been responsive to treatment, 80 percent (20 out of 25) had a syndrome characterized by overeating at night, insomnia, and no appetite in the morning. However, what the doctors described was binge-eating while these people were awake in the evenings or after an awakening during the night. They described patients who stayed up past midnight and ate high-calorie foods on at least half of all nights. But there was no such thing as polysomnography (or even sleep clinics) at the time, so it's impossible to know whether all of the patients described were actually awake during their nocturnal eating, or whether some were actually sleep-eating.

The patients had little or no success on weight-loss diets, so the doctors tried treating them with psychotherapy and "environmental manipulation" (such as locks on refrigerator doors). But only 2 of 20 patients (10 percent) with night-eating were able to lose more than a third of their initial excess weight—compared to 4 of 5 patients (80 percent) without nocturnal eating who were able to accomplish significant weight loss with the same treatment.

Even though this report brought up many interesting points and called for further investigation, only three reports on night-eating were published in the following 30 years. Two were single case studies, and none included polysomnography.

The first time sleepwalking was linked with night-eating in the medical literature was in 1981, when a doctor wrote a letter published in a medical journal describing a woman with schizoaffective disorder who had five episodes of psychotropic drug-induced sleepwalking in two weeks—and during these episodes, she "ate all the food available in the refrigerator." The following day, she had no memory of the events. She had a history of sleepwalking during adolescence, but not of sleep-eating.

The first clear-cut report on SRED was published in 1986. A 37-year-old man with a history of "rhythmic raiding of the refrigerator" spent six nights in a sleep lab and had 28 episodes of REM-stage sleep-related eating. He ate pork pies, potato chips, and biscuits, and drank bottles of soft drinks . . . and sunflower oil. Often, he'd manage to fall into stage 2 NREM sleep while still chewing his food, and have no memory of the episodes the next morning. He was a happily married man with no psychiatric disorders, and didn't pursue treatment.

In 1988, three cases of "somnambulistic bulimia" were reported in two women and a man who had nightly episodes of binge-eating, including strange foods like raw bacon.[1] In the sleep lab, arousals from slow-wave sleep with complex behaviors were documented in all three patients—classic for a non-REM sleep parasomnia.

Our center formally recognized this sleep disorder and named it in 1991. Besides sleepwalking, we identified sleep apnea, restless legs syndrome, and certain medications as triggers for SRED, along with an idiopathic group that had no identified trigger.

CHARACTERISTICS OF SRED

Sleep-related eating disorder usually arises from NREM sleep disorders like sleepwalking and sleep terrors, but it may happen at any time throughout the night—including during REM sleep. The person may get out of bed and raid the refrigerator or pantry, or may reach for whatever food or drink (or, sometimes, inedible substance) is near the bed. This is an involuntary action, though the person's level of consciousness and their subsequent memory of the event varies. Some people have no memory of these events at all and have to be convinced that they did indeed get up and eat, while others have some awareness but feel no control over their actions. There may be multiple episodes of SRED each night, or it may occur sporadically. Often, a person will have a long *prodromal* phase, which means that they'll have occasional sleep-eating behaviors while sleepwalking, which increases in frequency and severity over the years until they really become a problem. For many of these patients, within a matter of months, sleep-related eating becomes the exclusive or nearly exclusive sleepwalking behavior.

There is tension associated with the urge to eat, but the compulsion to eat is not associated with hunger, which is a notable and peculiar feature of SRED. Sleepers usually make a beeline straight for the kitchen, knocking over things in their path, but if they remember the experience later, they will rarely say that they felt hungry—just driven. They feel a great deal of frustration if they can't find or access food (such as if there is a lock put on the refrigerator). Sleeping in a hotel, for instance, can cause frenzied activity, like foraging through suitcases looking for something to eat. The drive to eat is often so strong and irresistible that even sufferers who get up in the night with a full bladder will nevertheless walk by the bathroom without hesitation and continue on their straight line to the kitchen.

Not only do sleep-eaters not feel hungry or thirsty as they're desperately searching for food, they typically do not have abdominal pain, nausea, heartburn, or hypoglycemic symptoms during their arousals. They may wake up in the morning feeling full and bloated, though, with no appetite for breakfast. It is interesting to note also that sleep-eating goes on irrespective of whether the person eats just before bedtime.

But what often disturbs people more than the fact that they sleep-eat is *what* they sleep-eat: very strange combinations of food, raw meat, condiments, or nonfood items like soap, hand cream, and buttered cigarettes! One of my patients used to eat salt sandwiches, or just pour salt into her hand and eat it. Another preferred sugar sandwiches or peanut butter, salt, and sugar sandwiches. Other reported food choices include cupcake frosting, mayonnaise, pasta, rice, chocolate, dog food, cat food, sour cream, coffee grounds, and ketchup in milk. The foods people consume during sleep-eating are generally not those they prefer to eat during wakeful hours.

Food preparation can be fairly complex, too—people may cook full meals, use blenders, toasters, or microwaves, and slice food (and sometimes their fingers) while sleepwalking. Generally, there's a messy component to SRED. Rarely do sleep-eaters clean up properly after their nighttime rituals, so they'll often find evidence of their actions the following day: food wrappers left out in the kitchen, scraps on the floor, crumbs in the bed, spills on the table, food mashed into the pillow. They might also leave the stove on or the refrigerator door open. They often eat with their hands right out of the refrigerator, in a fashion that would give Miss Manners "the vapors."

Other bad habits don't necessarily transfer to sleep, though. Very few people who smoke do so during sleep-eating episodes. And sleep-eaters virtually never drink alcohol, even those with a history of alcoholism.

The multidimensional toll that SRED takes on the millions of sufferers can be enormous, including the relentless sleep disruption, pervasive tiredness, progressive weight gain, and interpersonal stress. The weight gain can lead to obesity and the induction or aggravation of diabetes, hypertension, and elevated blood cholesterol and triglycerides, which in turn carry major risks for cardiac disorders and stroke. SRED can also cause tooth decay and lead to other serious dental problems.

MONTEL WILLIAMS AND SLEEP-EATING

On a talk show about sleepwalking, host Montel Williams revealed his own sleep-related eating behaviors: "I get up at 2 to 4 in the morning, every morning, and I'm eating something. And it's gotten to the point where I have to literally take raw foods out of my refrigerator because I have eaten some raw stuff. . . I know I did because I wake up in the morning and there's a pack of chicken and there's a bite missing out of it. I can take a whole pound of ham or bologna and eat the whole pound . . . and then wake up in the morning and not realize that I ate it and ask, 'Who ate my lunch meat?'"

Who Gets SRED?

About three-quarters of sleep-eaters are female, and the average age of onset for sleep-related eating is in the twenties.[2, 3] SRED can appear spontaneously, or be associated with sleepwalking, restless legs syndrome, obstructive sleep apnea, various other disorders, and medications.

Nocturnal Eating Syndrome and SRED

Nocturnal eating syndrome was first understood as excessive eating in the evening before falling asleep, but now we know that this disorder leads to eating episodes during the night as well. Nocturnal eating syndrome (NES) is a form of sleep-interruption insomnia. It shares some similarities with SRED in regards to the inappropriate timing of eating and the sleep disruption. NES is not a parasomnia nor a disorder of a "twilight state," but is strictly a disorder of wakefulness with full recall the next day. Also, NES is not associated with a primary sleep disorder, apart from insomnia. In contrast, SRED is often linked with other parasomnias. It is still not clear whether NES is a sleep-disruption disorder that causes eating, or whether sufferers have an eating disorder that leads them to disrupt their sleep to eat.[4] Morning anorexia (not eating breakfast) is common in NES (with both evening and middle-of-the-night overeating) and in SRED. Those with these disorders wake up feeling as if a rock is sitting in the pit of their stomach, and they have no desire to eat. They feel out of sorts—not eating when they should and eating when they shouldn't. And they often drag around during the day.

A person with NES may have very little appetite during the day but feel a strong need to eat late at night before bed. This may cause or aggravate insomnia, whether they act upon the urge or not: Eating large portions or snacks with sugar or caffeine can cause insomnia, and a strong feeling of hunger can make it very difficult to fall asleep.

Researchers have concluded that those with NES awaken throughout the night and experience difficulty going back to sleep (a condition called *sleep-maintenance insomnia* or *sleep-interruption insomnia*), unless they eat, and in the morning they remember what they did. With SRED, many patients hardly awaken while they eat, and in the morning they either don't remember or have hazy memories. There are many people with sleep-maintenance insomnia, and people use different techniques to get back to sleep. So, it's quite possible that NES sufferers represent a subgroup of sleep-maintenance insomniacs who find that eating helps them get back to sleep, whereas other people may prefer other remedies, such as reading, listening to music, drinking warm milk or tea, having sex, and so on. In other words, it may be that the insomnia comes first.

RISK FACTORS FOR SRED

- **Family history.** A family history of parasomnias—in particular SRED and sleepwalking—increases a person's risk of developing SRED.
- **History of other sleep disorders.** As is true with most sleep disorders, having one makes it more likely for you to develop others. In this case, sleepwalking, sleep terrors, RLS, PLMD, and OSA are particularly likely to be predisposing factors. In some cases we've seen, treating OSA with nasal CPAP therapy completely controls SRED.
- **History of Eating Disorders.** A preliminary study on the prevalence of SRED in various clinical populations found the highest rates among inpatients and outpatients with daytime eating disorders such as bulimia and anorexia nervosa.[5] However, these patients rarely present to sleep centers on account of SRED.
- **Quitting smoking.** In a published study of ours from 1993, two patients developed nightly SRED after quitting smoking cigarettes. We have seen more cases since then.
- **Quitting substance abuse.** From that same study, one man who had quit drinking after 19 years of alcohol abuse and one woman

who had quit drinking and using opiates and cocaine developed SRED. We have seen other similar cases and also cases triggered by stopping amphetamine abuse.

- **Experiencing stress.** In many cases, the onset of SRED can be linked to a time of extreme stress. We've seen patients whose SRED began after a parent's stroke or heart attack, after the birth of a child, after a relationship breakup, and so on. Often the stress involves separation anxiety. Unfortunately, the SRED behavior tends to stick around long after the stress is over. In one case, a 13-year-old boy developed sleep-eating while caring for an alcoholic father; his disorder persisted for 34 years.

- **Taking tricyclic antidepressants.** One of our patients, a 33-year-old woman with a 5-year history of SRED, realized that her night-eating began immediately after she began taking amitriptyline (a tricyclic antidepressant) for migraine headaches. Within one week of stopping the medication, her SRED remitted completely.

- **Taking other medications.** There are case reports describing various medications that have triggered SRED, including olanzapine (Zyprexa),[6] risperidone (Risperdal),[7] and zolpidem (Ambien).[8, 9] Discontinuing the medications has led to remittance of SRED behaviors.

Ambien and SRED

Zolpidem (Ambien) is the most commonly prescribed medication for insomnia. In 2002, doctors at the Mayo Clinic reported on five patients with RLS who were previously misdiagnosed with insomnia and treated with Ambien by their primary-care physicians. All five developed SRED after starting zolpidem and were always amnesic, meaning that these patients had no memory of eating during the night. When they stopped taking zolpidem and had their RLS and any other sleep disorder treated effectively, the SRED stopped. The doctors concluded that zolpidem may cause or worsen sleep-related eating, at least among people who have RLS and perhaps other sleep disorders that cause frequent arousals.

Three years later, our sleep center presented further evidence that zolpidem can cause amnesic SRED—even in patients who do not have RLS. We studied 19 cases of zolpidem-induced SRED over a two-year period in the sleep and psychiatry clinics at our hospital. (During the same time period, just one person was diagnosed with SRED that was induced by a different medication—triazolam.) Most of the people with SRED were women (84 percent), and 16 of 19 (84 percent) had recent or current major depression. Ninety percent (17 of 19) were also on antidepressant medication when the zolpidem-induced SRED began.

All of the patients had taken other sedative-hypnotics and other psychotropic drugs at bedtime, and didn't have sleep-eating as a side effect from any other medications. In other words, these patients served as their own "controls" in showing the specific Ambien–sleep-eating association. When they stopped taking zolpidem, the SRED stopped. Females with insomnia who are taking medication for major depression, who are taking 10 to 20 mg of zolpidem (medium-high dose range), and who may also have complex medical histories appear to be most vulnerable for zolpidem-induced SRED. We have now seen a total of more than 30 cases over a three-year period.

"RESCUE ME" EPISODE: TAKE-OFF ON THE AMBIEN STORY

In the highly rated firefighter series starring Dennis Leary on Fox TV, one episode was ripped directly from current news headlines about the hazards of Ambien misuse. A fireman is shown to mistakenly take four "wrong colored pills" that were actually the sleeping medicine "Sominol." He then engages in a prolonged sequence of highly inappropriate behaviors, including sleepwalking, sleep-driving, sleep-eating, sleep–food-shopping at a supermarket, and more peculiar activity. A cop refers to his behavior as "looney tunes." Dennis Leary's character, upon hearing about this bizarre Sominol effect in his comrade, says, "It's been in the news." So this reaction to Sominol was already part of common knowledge in "Rescue Me," similar to the current Ambien story in real-life.

CONSEQUENCES OF SRED

1. **Weight gain.** One of the major, obvious consequences of SRED is weight gain. In fact, almost half of our SRED patients are officially overweight, if not obese, using standard body mass index

criteria. The types of food people tend to eat during SRED episodes are high-carbohydrate, heavy foods like cakes, brownies, peanut butter, and ice cream. This behavior brings with it all the dangers of obesity, including increased risk for diabetes, high cholesterol, heart disease, and so on.

2. **Inability to lose weight.** On the other end of that continuum is an inability to sustain weight loss even when dieting. One patient with morbid obesity and OSA had to lose 18 kg to be eligible for gastric bypass surgery, but found that impossible because (in part) of her nightly sleep-eating.

3. **Social embarrassment.** Sleep-eaters are rarely careful and polite about their nightly table manners. They are often distressed to hear that they "pigged out" on ice cream in their underwear. One man says his girlfriend showed him that he had taken a large bite out of a frozen pizza in his sleep. His first concern was, *What must she think of me?* Those with SRED are often worried about sleeping in unfamiliar surroundings, going away to college, or starting intimate relationships.

4. **Shame.** Many people with SRED describe a feeling of shame because of their loss of control, which they sometimes attribute to a lack of self-discipline or willpower. They may experience lowered self-esteem and depression because of their night-eating habits.

5. **Disregard of food allergies.** Sleep-eaters may have no concern for their food allergies or sensitivities while in this compromised state. A person with an allergy to chocolate or peanuts may happily create a peanut-butter-and-chocolate-syrup sandwich during sleep. A person with lactose intolerance may chug whole milk. Some patients have to keep these types of foods out of the house, even if family members don't share the food sensitivities.

6. **Potential for injury.** There is risk of injury from food preparation (cutting, cooking, leaving appliances on, drinking boiling liquids) and from choking, especially if the person eats or drinks while lying down in bed.

7. **Dental complications.** Tooth decay and other dental problems can occur, at times with extensive damage that can be quite costly to treat.

8. **Dangers of medication.** Certain medications, particularly mono-

amine oxidase inhibitors (MAOIs), a class of antidepressants prescribed for depression and anxiety, carry strict dietary restrictions.
It's especially dangerous for a doctor to prescribe an MAOI to a
person with sleep-eating tendencies.

9. **Daytime tiredness and sleepiness.** Those with SRED may have
highly fragmented sleep, and as such, they may not feel restored
the next morning and may be tired or sleepy throughout much of
the day.

ODD STATES OF CONSCIOUSNESS

The level of consciousness during episodes of SRED, as well as the extent of subsequent recall of the episodes, can vary widely across patients
with SRED, but generally there is not much variation for each affected
person across multiple episodes. About half of patients have partial awareness during their episodes and afterward have spotty recall, with another
third of patients being completely unaware of their actions and having no
recollection of their episodes. The remaining one-sixth of patients have
considerable awareness (but no control) during an episode, and have recall
of the episode on the following day.

Among the patients I've worked with, there was a 42-year-old woman
with 20 years of nightly sleep-eating who could recall some episodes in
which she would decide *not* to eat a cat-food sandwich that she had just
prepared in the middle of the night—not because she suddenly realized
(with any horror) that it was actually cat food she was about to consume,
but because in her mind at the time, she was not inclined to eat a sandwich. Another patient kept trying to find a "safe place" for a head of let-
tuce that she was carrying around while sleepwalking, and eventually she
hid it in the bathroom.

Carol's Story: Sleep(less)Walking

I slept just fine until I was 31 years old. Then something strange happened: I couldn't keep my legs still at night. I didn't have any itching
or pain—I just could not stop moving my legs. In the beginning, I
could get about an hour and a half of sleep before my legs would

wake me up, but it got worse and worse, until I was sometimes sleeping just fifteen minutes a night.

I had to keep moving. I'd turn myself upside down so my legs were up where my head should be on the bed. I didn't want to wake up my husband, so I'd walk out of the bedroom and pace around the house. I'd think, maybe I could get some sleep in the bathtub, or in the car, or in the closet. It sounds crazy, but you'll try anything when you're that desperate.

And in the middle of the night, I'd eat whatever sweets I could find. Some days I'd think I did okay, but then I'd find candy-bar wrappers or cookie crumbs all over the kitchen counter. I wouldn't even remember that I had eaten.

I couldn't travel much because pacing around in a hotel room all night just didn't work. I remember going camping once and being so frantic when everyone else woke up for the day and I hadn't slept yet. *They can't get up yet!* I thought. *I still need to sleep!*

My husband always wanted to open a restaurant in our small town. He was so hyped up and energized, so I said okay, even though I thought, *I can't do this!* We were so busy. I did all the baking. I was up there by 5:00 in the morning and I worked all day. I remember putting my head down on the basement steps on a piece of cardboard. I don't know how I functioned on such little sleep. I read that there are people with sleep disorders who shouldn't be driving, and they're right—I can't say I was any safer to be on the road than somebody with sleep apnea.

I went to see doctors about this many times, and they just put me on medicine to knock me out, but it didn't work. They didn't seem to understand about my legs. Maybe they never heard of such a thing back then, so I don't really blame them. I would tell people about it and they'd say, "Oh, I get that, too," but they didn't understand. Maybe they couldn't sleep for 15 minutes because their legs were restless, but they didn't have what I had. I thought no one did.

There are so many people out there who are handicapped or disabled or living in pain, I told myself. *There are people so much worse off than I am.* Yet when you have something like this, it's sort of a big thing to you. I won't lie . . . I thought about suicide. I wouldn't do it, but I thought about it. I don't know how I got through it. I don't think I could do it again.

I'm 71 now, and I finally got help 16 years ago in October when I went to see Dr. Schenck. I'll never forget that day. I described my symptoms just the same as I had described them to everyone else—there isn't much to tell. I just can't keep my legs still and it keeps me up all night long, and I eat during the night. But he was the first person who seemed to understand. "I think I know what this is, and I can fix it," he told me. I was there in my pajamas and we did an overnight study and he told me I had restless legs disorder and sleep-related eating disorder.

He prescribed a medicine for me, and I remember my husband saying that we would get the prescription filled on Monday, because this was the weekend. Dr. Schenck said, "No, I'm sending you home with this medicine today." And I took it, and that night, I slept for eight hours straight. It was wonderful.

I sleep eight to nine hours a night now. I can't sleep during the day. I have to wait until I take my medication at night. My whole life changed when I got my disorder under control. Dr. Schenck invited me to talk about it, so I was interviewed for a Minnesota newspaper article, and I was on *Good Morning America* and in *Reader's Digest*. I remember the phone ringing nonstop after that article came out. Every time I put down the phone, it would ring again, from people all over the country and even from other countries. They all had these same kinds of problems and thought they were the only ones, too.

So now I think, if I can just get my story out to one doctor who could help one person like me, that would be worth it.

Carol's Treatment

Carol is a perfect example of someone with a "parasomnia shadow." That is, her fairly common sleep disorder led to a more extreme parasomnia.

My initial approach for treating Carol's dual sleep disorders was to try to control the presumed underlying disorder (RLS) and then see what would happen to her sleep-eating disorder. The results were excellent, as standard treatment of severe RLS immediately controlled both her RLS and the SRED.

As Carol's story mentioned, in 1992, *Reader's Digest* published an article about sleep disorders, focusing mainly on our sleep-disorder center, and

Carol was one of the patients featured in that article. Interestingly, the writer focused only on Carol's RLS, not mentioning her SRED. So 14 years ago, SRED was considered "beyond the fringe" and not on the radar screen, which, of course, is no longer the case.

CONDITIONS THAT CAN LEAD TO SRED

Sleepwalking

Restless legs syndrome

Obstructive sleep apnea

Medications (zolpidem [Ambien], olanzapine [Zyprexa], risperidone [Risperdal], lithium, tricyclics, anticholinergics, others)

Nocturnal dissociative disorder (with night-eating personality; see chapter 14)

Eating disorder (bulimia nervosa; anorexia nervosa)

Idiopathic (sui generis)

TREATMENTS FOR SRED

We've studied the efficacy of many potential treatments for SRED, and found that only medications and medical therapy for sleep disorders (such as the nasal CPAP for OSA) are generally effective.

Hypnotherapy and self-hypnosis, biofeedback, acupuncture, and psychotherapy had no benefit on the sleep-eating habits of the patients we've studied. Though in one case report, a therapist used a woman's phobia as a way to combat the disorder. This 36-year-old woman with SRED was unresponsive to medication and hypnosis, and had been eating raw meat, butter, and uncooked vegetables during sleepwalking. Because she had a phobia for snakes, her therapist told her to place a toy snake in the kitchen every night, which put an end to her refrigerator-raiding habits.[10] However, this is risky therapy. In another patient, the snake phobia could have caused a major panic because of the person seeing it at night during this vulnerable state.

Some patients have put locks on their refrigerator doors or bedroom doors—but the urge to eat is so compelling that people may sustain injuries trying to break through the locks. In a *Washington Post* article, David Neubauer, associate director of the Sleep Disorders Center at Johns Hopkins Medical Institutions, described a patient who managed to break the door frame of her bedroom after her mother locked the door (on the patient's request) to get her to stop sleep-eating.[11] Other behavioral strategies people have used include putting notes on the refrigerator door (often insulting ones, like, "Don't eat, you pig!"), asking family members or roommates to sleep in the path between the sleep-eater's bedroom and the kitchen, or "punishing" oneself for sleep-eating by having to do something undesirable the next day. Sometimes these behavioral strategies help for a period of time, but most often they aren't effective.

Topiramate (Topamax), a rather new anticonvulsant medication, is currently the treatment of choice for SRED associated with sleepwalking and for idiopathic SRED. Over a 30-month period, I prescribed topiramate to 17 patients with SRED. When I followed up with the patients six months later or more, 11 of them (about two-thirds) reported complete or substantial relief from SRED episodes. All 11 also reported that sleep became restorative, and they all lost weight (the primary outcome measure of the study)—an average of 21 pounds. Tingling sensations (*paresthesias*) are a possible side effect of the drug, but are usually tolerable.[12] Similar findings have been reported from another center.[13]

Sometimes topiramate controls the eating but does not control awakenings. In these cases, I prescribe low doses of clonazepam or trazodone to "smooth out" the sleep.

Bupropion (Wellbutrin) or fluoxetine (Prozac) may also be prescribed in some cases, particularly when major depression occurs concurrently with the sleep disorder. These may be prescribed alone or in conjunction with other medications to control SRED.

THE STIGMA CONTINUES

Syndicated newspaper columnist Garrison Keillor published a column about his experiences with sleep-eating in February 2006.[14] It appeared in many newspapers, and on Salon.com as "Confessions of a Sleep Eater." He explained that he ate frozen waffles, an entire quart of ice cream, chicken nuggets, anchovies, and more in his sleep. He says he once awoke to find a dozen White Castle double-cheeseburger cartons under his pillow, then saw his car keys and drive-through receipt on the floor, and found his car in the driveway with the driver's-side door wide open. He deduced that his sleep-eating had led him to drive in his sleep—quite a dangerous problem!

Granted, Keillor is a humorist, so I wasn't expecting a detailed write-up of resources for sleep-eaters, but my colleagues and I were flabbergasted by his concluding lines: "The way to overcome sleep-eating is to face up to the truth about yourself. I did it and so should you. Has anyone told you that you could stand to lose a few pounds? Well, you could."

My colleagues and I sent a letter to the editor of our local newspaper, expressing our dismay at this simplistic and possibly harmful advice. Nowhere in the article did Keillor explain what "the truth" was that freed him from this disorder. Certainly, there are factors to consider that are within a person's control when evaluating a sleep disorder. If you have a wildly fluctuating sleep schedule, or if you know you have a sleep-eating problem and you keep lots of junk food next to your bed anyway, for example, you may be exacerbating the problem. But most often there is no psychological "truth" to face in order to end the amnesic sleep-related eating and sleep-driving. If you correct bad habits, or if you're doing all the right things and still have a problem, it's time to stop looking inward and seek outside help.

Keillor's advice puts the onus on the sufferer, as if all that's necessary is to take responsibility for bad behavior—as if, in other words, sleep-eaters are just bringing on their own problems and can simply choose to stop.

This is very unfortunate because SRED is already such a guilt-provoking topic, and it doesn't need to be. SRED is a complex disorder that often evolves from another sleep disorder and can be brought on by stress or medications. Those with persistent SRED need to be evaluated by sleep professionals, and can be treated effectively—no guilt required.

THE FUTURE FOR SRED

Because of the increasing media attention to this disorder, public awareness about it is increasing. However, many primary-care physicians still fail to ask questions about sleep during routine exams, and thus can easily miss the presence of parasomnias, including SRED. Especially when there's a patient who gains weight or has trouble losing weight despite doing the "right" things (weight-loss diet and exercise), it's a good idea for doctors to ask about nocturnal eating habits, morning anorexia, and possible sleep-related eating. Often, those with this disorder are unaware of their behaviors until someone in the household points them out—or after there are arguments over who ate the last piece of pie or the entire pound of cheese, and no one owns up to it.

SRED is a challenging condition to treat, but more than 60 percent of patients achieve substantial and lasting control with specific medications or with treatments for underlying sleep disorders. Future research may focus on the similarities and differences between night-eating syndrome and SRED, and answer further questions about the phenomenon of abnormal eating during the night: To what extent is it a sleep disorder versus an eating disorder? What other medical disorders or medications can trigger SRED?

Those with SRED behaviors should be aware that they're not just "weak-willed." This is a sleep disorder over which the affected person has no control; it's nothing shameful, and effective treatments are available.

Chapter 13

REM SLEEP BEHAVIOR DISORDER

*Bizarre and Violent
Dream Enactment*

Sleep is supposed to be a time to rest, relax, recharge—to leave behind the cares of the day, and perhaps also a time to enjoy some dreams as an added bonus. But what if you find out that you become transformed in your sleep from a peaceful person to an attacking monster? A bit like in werewolf mythology, people with REM sleep behavior disorder, or RBD, seem to become wholly different creatures in their sleep.

For most people, there are fairly clean distinctions between NREM and REM sleep. REM sleep is when most dreaming occurs. The brain is highly active and the eyes dart back and forth rapidly, but the body is immobile. This virtual absence of skeletal muscle activity is called *atonia*.

Then there are those with RBD, for whom the "rules" of sleep don't apply . . . with frightening consequences.

Those afflicted with RBD lack the protective mechanism of REM-atonia that immobilizes them during REM sleep. Therefore, their bodies are free to get up and act out their dreams. This scenario is made worse because RBD dream-enactment does not involve one's average dreams—they're pointedly more vivid, intense, physically active, confrontational, aggressive, and violent, with classic fight-or-flight scenarios.

That's what makes RBD such a terrifying problem: Those with the disorder are prone to attack or to try to kill intruders and fight off wild animals in their sleep . . . except that those "intruders" and "wild animals" are usually their bed partners. They may yell and shout profanities, and they are at a high risk of injuring themselves or their spouses, even fatally.

THE DISCOVERY OF RBD

At the start of my career as a sleep doctor in 1982, the second patient I saw for an evaluation turned out to be our first dream-enacting RBD patient—Donald Dorff, whom I mentioned at the beginning of this book. Don's case challenged much conventional wisdom about sleep, and has paved the way to understanding more about the broad range of dissociated states of sleep and wakefulness, the "mixed states of being" that afflict many people.

Don was recently retired, 67 years old and playing golf five days a week year-round (in Minnesota and Florida). He came to see me after gashing his forehead one night in the midst of what he called a "violent moving nightmare." This was a new problem that had started after his retirement, and he was not happy. His sleep problem was a nuisance and also becoming dangerous. His regular doctor and a psychiatrist had found nothing wrong with him, so they sent him to us for an evaluation. Don Dorff, my second patient, became our center's first documented dream-actor, our index case of RBD.

In a memorable quotation from a *National Geographic* article describing his case, Don said that one night he was playing football in his dream, and was carrying the ball through the line of scrimmage as a running back when "there was a 280-pound tackle waiting for me, so I decided to give him my shoulder. When I came to, I was on the floor in my bedroom. I had smashed into the dresser and knocked everything off it and broke the mirror and just made one heck of a mess. It was 1:30 a.m."

When Don first came to see me, we had never seen anything quite like his problem. In the sleep laboratory, we hooked Don up all night with electrodes monitoring his vital functions. Don's sleep architecture looked pretty normal—he cycled normally across the different sleep stages. But he had some episodes where he abruptly threw punches, then quickly lay still again. Our night technician noticed something else strange about these events. His EEG brain-wave activity and rapid eye movements showed that Don wasn't awake, nor had he just been aroused from NREM sleep, as would be seen with a violent sleep terror or confusional arousal.

His EEG showed the low voltage and fast frequency wave pattern of REM sleep, but his chin, leg, and arm muscle sensors showed twitching and prolonged periods of muscle tone. In normal REM sleep, the muscles are supposed to be inactive, with barely a twitch and virtually no tone. So Don was demonstrating distinctly abnormal REM sleep with violent behavior. It is quite possible that our technician thought the equipment was malfunctioning at the time, but Don's movements were real. And he recalled dreaming during the episodes in the sleep lab.

The next morning, Mark Mahowald and Andrea Patterson, our sleep-center director and chief technologist, examined and reexamined his videotape and his recorded polysomnographic data and realized that Don demonstrated a dramatic loss of the usual—and protective—atonia that occurs during REM sleep in all mammals. Don was thus released from the paralytic shackles of REM sleep, and was able to carry out dreams—but not his customary dreams. His dreams had become radically transformed and disturbing since he developed his sleep problems.

The dream plots were almost all confrontations with unfamiliar people or animals, where Don had to fight or escape in order to survive. Sometimes the dreams involved sports, but in a violent manner, such as playing American football. The dreams were so distressing that Don called them "violent moving nightmares." He came to us needing help in controlling his excessive sleep movements, his sleep violence, and his nightmares.

As a child, Don didn't have any sleep problems. He shared a small bedroom with three brothers uneventfully. On his wedding night, however, his wife, Marie, reported being scared by his strange habits: sleep-talking, tooth-grinding, groaning, and minor body movements. And this went on most nights for 41 years before it developed into anything more serious. When he was 63 years old, he had his first "physically moving dream," and these dreams soon became frequent, violent, and unwanted intruders into his sleep.

Don described a dream where he was on a motorcycle, getting into a fight with another motorcyclist as they rode down the highway. The other man tried to ram into Don's motorcycle, so he kicked the other motorcycle away—except that he was actually "kicking the hell out of" his wife, who was rather startled out of her own sleep.

He frequently fell out of bed, slammed into dressers and lamps, and punched and kicked his wife. The next day, he'd discover black-and-blue bruises and cuts.

Treating Don

Because we'd never seen anything like this before, we didn't know how to treat Don—so we employed a bit of trial and error. My first thought was to put him on a tricyclic antidepressant, which reduces the total amount of REM sleep. I hoped that if Don spent less time in the REM stage, he'd have fewer symptoms—but that didn't work.

We noticed that he also twitched and kicked his legs during NREM sleep, so I next decided to treat that with the medication commonly used at the time to treat periodic limb movements during sleep: clonazepam. I hoped that, because we knew that clonazepam was effective for treating one sleep-related movement disorder, it might help another one, too. It was immediately successful; Don's RBD went into remission, and he reported that the disturbing dreams stopped, too.

Within three years, we saw four other patients who had the same symptoms and sleep-lab findings that Don did, so we knew it wasn't just a strange one-time occurrence. They, too, were having violent behaviors during sleep in association with vivid, aggressive dreams; and they, too, showed in the sleep lab that they were perfectly able to move and to act out dreams during REM sleep when they were supposed to be immobile. After a total of ten patients with this same strange condition had been gathered over a five-year period, we realized that it was possible that there were many others out there with a parasomnia that we named the *rapid-eye-movement (REM) sleep behavior disorder* in 1987. After our publication in the *Journal of the American Medical Association,* the media response was so great that many people with RBD and other parasomnias started coming to our sleep center, confirming that parasomnias are not uncommon conditions.

MARTIN AND GERTRUDE

In 1990, a happily married couple came to our clinic after their primary doctor referred the 69-year-old husband to me. I interviewed the couple and reviewed their completed sleep-center questionnaire, and immediately suspected RBD by the way they answered questions. His chief complaint was "active dreaming—arm and leg movements, acting out dreams, occasionally falling out or jumping out of bed."

The following accounts are in their own words.

Gertrude's Story

I remember exactly the month Martin's sleep disorder started— February of 1983—and the reason I remember is that it was the month he retired. We wondered if there was a connection, but I don't think so. The best I can describe is that he would literally go wild at night. It started mostly as a combination of kicking violently in bed and thrashing his arms. He would throw his arms all over when he went through these episodes. Too often, he would do so much real loud screaming, and that was very disturbing. I'd wonder why he, why *anyone,* would scream so loud at night. Rarely was it words that you could understand—mostly it was nonsense talk. I think the screaming was the most disturbing part to me. It was violent screaming, and it would wake me out of a sound sleep. Afterward, I couldn't stop shaking for a long time.

One of our daughters got divorced and was living with us for four months with her two little boys, and I remember the six-year-old would say, "Grandpa, why do you scream so much at night?" It must have been pretty traumatic for the little boys.

When his arms were flailing, sometimes he'd hit me over the head, and I was actually scratched a few times, but there were two really dramatic episodes that neither one of us will ever forget. He grabbed my hair one time, pulling it for all it's worth, and I woke him up to ask him what was going on. He said, "There's a skunk in the tent. I'm getting rid of it." Earlier, he had gone fishing in Canada, and actually had a skunk in the tent, and he kicked it out. The scariest of all was the time he had his hands around my throat and was actually choking me. That, of course, is something I'd never forget because that feeling of having pressure on your neck is terrifying.

It was never easy to wake him up. I could get him to stop, but it was hard to get him fully awake. He would go right back to dreaming in no time.

But he was always trying to protect me. He thought somebody was trying to get me, people trying to get in our house. He thought they were coming up the stairs, and he jumped out of bed a few times. He would hit a piece of furniture and end up with black and blue spots. We had a bay window in our bedroom, and one of our daughters who lived with us used to worry that her dad would jump out the window, which would have been possible. He would've fallen out on the cement driveway.

I remember he would dream approximately four times a night. The longest dreams, and the most violent with the most kicking and real loud screaming, usually were closer to morning. Once in a while, they were maybe 1 or 2 o'clock, but more of them were between 4 and 5 a.m. We were traveling quite a bit at that time, and the thing that concerned me was, if he did this loud screaming and we were in a motel, somebody would alert the authorities or the officials of the hotel that I was being abused. It really changed the way we traveled. We never did it without apprehension.

I never dared face him in my sleep, because he flailed his arms and I would probably get socked in the eyes. He used a lot of force—that's the thing that was so scary. The other thing I worried so much about was that he did have guns for hunting. I was so afraid that he would get up and get a gun and probably shoot us, so I talked him into selling his guns. I never did hide the knives, but when he was having his strongest episodes, I worried about it.

I still have the articles I saved from the 1980s, where I read about RBD for the first time. One woman said, "One night I felt something around my neck and opened my eyes and saw that he was strangling me." I had exactly the same thing happen to me! So in 1990, we went to our family doctor because Martin was concerned over this. Of course, the family doctor had never heard of anything similar, but my husband said, "My wife has articles about Dr. Schenck, and his patients' problems sound so much like mine." So we got a referral to Dr. Schenck and then Martin went for an overnight sleep study at the Hennepin Medical Center. They said his legs kicked 47 times in one hour, and that he had REM sleep behavior disorder and restless legs syndrome.

Since he's been treated, it's a whole new world. He still kicks, but I don't believe he's ever had a violent episode since then. He doesn't do loud screaming, but every now and then he'll give a shout like he wants to stop somebody—like kids are aiming a ball at the house or something, and he'll yell at them to stop that. Occasionally, he'll call out my name and say, "What do you want?" When he's napping in a chair, his lips move like he's saying words, even if he doesn't say anything. He naps in the chair often, and if you watch, he's talking all the time, moving the lips. I've timed him as long as 30 minutes in one dream. I can see his eyes moving back and forth during his dreams as long as there's enough light in the room.

We quit sleeping together about the mid-nineties, because I still couldn't get decent sleep. They say that the spouses have a harder time than the people with the sleep disorder. I do sleep with him on occasion in motels or when staying with relatives.

It's been 16 years since he's had help. He's had a worthwhile 16 years. I try to remember those seven years when he didn't have help, and I even asked my daughter what she remembers, and she said the screaming would wake her up and she would be very shaken for a long time. I think he still dreams just as much, but the medication controls the terrible part. The violence is gone. We were able to go back to traveling and feel like I wasn't afraid that people would call the cops to say I was being abused. It was a lifesaver. There's no other way to say it.

Martin's Story

A lot of my dreams were with animals. They were realistic dreams, but I couldn't remember them by the morning. If my wife comes in and says, "What are you dreaming?," I might be able to tell her, but I'll forget by the morning. When I choked my wife, I think I was dreaming about getting rid of an animal.

The dream I remember best is one where I dreamed I was in King Arthur's court, and I was running through one of those long log houses and jumping over people with a sword as I ran. When I yell, I'm usually scolding people in my dreams. Sometimes my wife will ask what I'm yelling about, and I'll tell her, "I didn't like what they were doing, so I was scolding those kids." But it's not only kids. I think I yell at adults in large groups, too.

I fell out of bed sometimes, and the space was narrow, so I'd usually hit something and get kind of wedged in. I got plenty of black and blue bruises.

I served in the armed forces during World War II, terrible experiences. A lot of men have nightmares after that. That's one thing I've never had—maybe once I told my wife I had a dream about the war, but it wasn't violent.

I've been on Klonopin and Sinemet, and now they've added Requip, which is the new one for restless legs. My short-term memory and my handwriting have been affected. Dr. Schenck checks me regularly for Parkinson's disease, but he doesn't see any signs of it.

When I forget to take my medication, I spend the whole night pacing around the house, going upstairs, downstairs to do some painting, fixing stuff. I get very irritable and don't sleep, and in the morning, I think, "Oh, my aching back! I was supposed to take my pills!" My wife has a sister about 200 miles away, and I forgot to take my pills with us when we visited. Nothing I can do about it because they're not there. It's torture, but it shows what a difference the medications make.

The restless legs have gotten much worse over the years, and I've had to increase the dose of my medicine. We went to a movie late Saturday afternoon, and as soon as I got in the theater, my leg was starting to jerk around. That was an hour before I was supposed to take my medication. I walked out of the movie at the right hour, took my medication, and went back and didn't have any more jerking. So the medication really works. The other place where I can't sit still is church—that's terrible! Whenever I'm confined, that's when it's the hardest situation, where you can't get up without everybody noticing.

My wife has to be very alert to my medication. It takes two people to keep this under control. She is my guardian angel.

Love Conquers All

At the time I write this, Martin Ziebell is now 84 years old, and he and his wife, Gertrude, have been married for more than 56 years. When I last spoke with them on the phone, they were cuddled up on the couch by the fireplace on a winter afternoon.

One of the things that strikes me and is so touching about my patients with RBD is that so many of them have such strong marriages. Despite years of brutally dangerous sleep behaviors—including one event where he could have accidentally killed her—their love was never in question. She knew that who he was when he dreamed was not "really him."

"THE MINNESOTA DISEASE"

For years, I would run into other sleep doctors at the national meetings who jokingly referred to RBD as "the Minnesota disease" because we were the only ones who seemed to see it. It's pretty easy to miss or misinterpret the results of a PSG study that indicates RBD if you've never encountered it before, though, and don't know what to look for. It wasn't that no one with RBD had ever gone to another sleep clinic; it's just that other sleep doctors weren't recognizing the disorder. Some of these doctors mentioned to me that they thought these older men were waking up in the sleep lab and acting in a confused and belligerent manner as if they had Alzheimer's disease or some other dementia. But now, RBD is a well-recognized parasomnia with several thousand cases having been published in the medical literature worldwide. Symposia and group discussions pertaining to RBD are organized at virtually every national and international sleep meeting nowadays.

WHY AREN'T WE PARALYZED IN NREM SLEEP?

We don't need to be paralyzed during NREM sleep. We still have muscle tone, but we normally have no urge to move. Our brains are fairly inactive, so we aren't inclined to respond to stimuli. In REM sleep, in contrast, our brains are highly active and hallucinating, creating all sorts of "movies in our minds" that feature us as the primary actors. Thus, we have an inclination to move—and concurrently we have a protective mechanism to stop us from doing so.

BEFORE IT HAD A NAME

RBD certainly existed before we discovered and named it. Our patients have told us that before they received a correct diagnosis, they were often told they had psychiatric disorders (such as nocturnal psychosis, repressed aggression, or nocturnal post-traumatic stress) or else nocturnal seizures. Knowing what we now know about RBD, it's frightening to think about all the people who've been undiagnosed or misdiagnosed and untreated or treated with inappropriate medications.

There were signs long ago that RBD could exist, starting with a cat experiment.

In 1965, a group of scientists in Lyon, France, led by the renowned Michel Jouvet, was trying to locate the part of the brain responsible for generating REM sleep. They made lesions in 35 cats' brains to see if they could eliminate the mechanism responsible for REM sleep. When they made specifically placed lesions within an area of the brain stem called the *pons,* they found that the cats still went into REM sleep—but they lost the atonia that's supposed to accompany it. They reported that the cats showed strange behaviors during clear-cut REM sleep, such as "seeming to fight imaginary enemies or to play with an absent mouse, striking out with forelimbs and manifesting fear reactions."[1]

Another study, led by Adrian R. Morrison at the University of Pennsylvania, replicated these results and added to them.[1] During REM sleep, the cats in this study displayed hallucinatory-type behaviors. But depending on the size and location of the lesions, the cats' behavior during REM sleep differed, falling into one of four categories: 1) small twitching and jerking in the limbs, 2) exploratory behavior such as staring, head raising, reaching, grasping, and searching, 3) stalking and attacking behavior, and 4) locomotion, including running. These findings are mirrored in human RBD.

REM sleep is known by several names in the scientific community. It is sometimes known as *active sleep* because of the highly activated brain state during REM sleep. It is also known as *paradoxical sleep* because of the paradox between a highly active brain and an immobile body. The scientists working on the cat studies referred to the condition they created as *paradoxical sleep without atonia.*

These studies helped us to understand Donald Dorff and subsequent patients. In humans, however, lesions within the pons are rarely implicated in RBD, indicating that the causes of RBD in humans are much more complex than in cats.

DON QUIXOTE'S DREAMS

My colleagues presented me with a gift in December 1996, when I lectured in Barcelona at the annual congress of the Spanish Neurological Society: a copy of *Don Quixote de La Mancha,* the Miguel de Cervantes classic novel from 1605 that inspired the musical *Man of La Mancha.* They had marked a special passage for me in the "original" Castilian Spanish edition. A standard English translation describes the scene:

> He was in his shirt, which was not long enough in front to cover his thighs completely and was six fingers shorter behind; his legs were very long and lean, covered with hair, and anything but clean; on his head he had a little greasy red cap that belonged to the host; round his left arm he had rolled the blanket of the bed, to which Sancho, for reasons best known to himself, owed a grudge; and in his right hand he held his unsheathed sword, with which he was slashing about on all sides, uttering exclamations as if he were actually fighting some giant: and the best of it was his eyes were not open, for he was fast asleep, and dreaming that he was doing battle with the giant. For his imagination was so wrought upon by the adventure he was going to accomplish, that it made him dream he had already reached the kingdom of Micomicon, and was engaged in combat with his enemy; and believing he was lying on the giant, he had given so many sword cuts to the skins that the whole room was full of wine. On seeing this the landlord was so enraged that he fell on Don Quixote, and with his clenched fist began to pummel him in such a way that, if Cardenio and the curate had not dragged him off, he would have brought the war of the giant to an end. But in spite of all the poor gentleman never woke until the barber brought a great pot of cold water from the well

and flung it with one dash all over his body, on which Don Quixote woke up, but not so completely as to understand what was the matter.

Indeed, Cervantes had described an excellent case study of RBD, nearly 400 years before it was discovered and named. Just like Don Dorff, Don Quixote was dreaming of violent battles—and freely acting out these dreams.

Characteristics of RBD

RBD is a disorder not just of REM sleep but of all three basic states of existence: REM sleep, NREM sleep, and wakefulness. It is characterized by aggression and violence during sleep and not by any other kinds of instinctual behavior, such as eating or sex.

For RBD to exist, not only does the normal muscle atonia of REM sleep have to be compromised, but there must be excessive activity of the *motor pattern generators* in the brain stem, as identified in animal experiments. In other words, there must be a strong drive to move and behave while dreaming, and not just a release from the paralysis.

Interestingly, men with RBD usually don't feel tired in the morning, and daytime nap studies (i.e., the Multiple Sleep Latency Test previously described) don't reveal any excessive sleepiness, unless narcolepsy is also diagnosed. Although the reason for why RBD does not cause morning tiredness is not well understood (and in fact most people continue to feel rested from sleep despite having vigorous RBD episodes), a likely explanation is as follows: in published studies, most middle-aged and older men with RBD have excessive amounts of slow-wave sleep and "delta EEG power" that is considerably greater than what is found in comparably aged men without RBD. It is known that slow-wave sleep (a stage with predominant delta EEG power) features energy conservation. Therefore, we speculate that during sleep in RBD, the energy conservation associated with increased amounts of slow-wave sleep compensates for the increased energy expenditure associated with RBD episodes. The end result is feeling in "energy balance" and restored from sleep. In contrast, it is usually the wife who feels "wiped out" in the morning from the interrupted sleep caused by the husband's RBD.

There is a twist to this story: Although it may seem logical to conclude that the increased slow-wave sleep and delta power are an adaptive and "homeostatic" response to the increased energy expended during REM sleep in RBD, this is probably wrong. The available evidence strongly indicates that this association is primarily a reflection of global brain dysfunction across non-REM and REM sleep and also wakefulness in RBD. However, the good news for men with RBD, and their wives, is that something that is pathological may also serve an adaptive purpose.

In the motor system, there are typically periodic and non-periodic limb movements that occur without arousals—true within-sleep behaviors and not arousal behaviors. People with RBD have an increase in their percentage of slow-wave sleep (and delta EEG power), which may be a subtle indicator of brain dysfunction, but which also means that more energy is conserved to make up for the increased energy use during REM sleep. (Delta NREM sleep is an energy-conservation sleep stage). So the (happy) end result is a sense of restorative sleep when the person arises in the morning, despite all that moving and thrashing around during dreams. It is usually the spouse who feels "wiped out" in the morning from the interrupted sleep caused by RBD.

Most people with RBD do remember the dreams that occurred with their behavioral episodes, but at least 10 percent of people with RBD don't. Also, in a small subgroup of patients in our series, we found that atonia would kick in at some point during their dream-enacting behaviors. Their dreams would then alter to fit this new reality: The person would find themselves stuck in mud or trapped in deep snow, or believe they had fallen and couldn't get up.

Those with RBD are not more aggressive or violent than other people during the day, even though their dreams are significantly more aggressive, as shown in a recent study from Italy.[2] Anecdotally, the wives of men with RBD, who have been married for decades, often describe their husbands as particularly gentle and peaceful during wakeful hours, which is what makes it so difficult for them to understand how these men could become so vicious during sleep.

What typically brings them into a sleep clinic is injury to themselves or their bed partners (or both); 79 percent of the patients in our ongoing series came in with that as their primary complaint. The injuries typically involved bruises, lacerations, and fractures. Some of the more severe injuries we've seen are spinal injuries and subdural hematomas (blood collected on the surface of the brain, which may be deadly).

There are several other disorders that can look like RBD, with dream-enacting behaviors. It's important to conclusively prove RBD with PSG testing before treating it because the treatment for RBD can actually worsen other sleep disorders that may be mistaken for RBD, such as sleep apnea and dissociative disorders.

Unlike other sleep disorders, the hallmarks of RBD do show up virtually every night, assuming that a sufficient amount of REM sleep is present, even if there is no major episode—so one PSG study is usually enough to determine the diagnosis.

Altered Dreams

Most patients with RBD report that their dreams changed when the disorder began. The RBD dreams are vivid and action-filled, and are often interpreted as terrible nightmares. Rarely is the dreamer the aggressor in these dreams—he (or she) is almost always reacting to a perceived attacker. The attacker may be a person, animal, or insect, or an event like a fire (people and animals as attackers are the most common dreams). Some dreams are sports-related, where the "attacker" is an opposing team member. The dreamer has a sense of fear or anger and will fight back or escape.

DEMOGRAPHICS OF RBD

RBD primarily and overwhelmingly affects men over the age of 50—more than 85 percent of those with RBD fall into this category. To date, no one has been able to explain this male predominance. Although uncommon, RBD may also affect women and even children.

No one knows the exact prevalence of RBD, but it's been suggested that prevalence is approximately 0.5 percent of the general population. And it may be that more than 1 percent of men over the age of 50 years—the highest risk group—have RBD.

THE PROGRESSION OF RBD

Few people with RBD have histories of sleepwalking or night terrors in childhood. But about 25 percent have a prodromal period—sometimes very lengthy—where some behavioral symptoms such as vocalizations and movements are present during sleep before the disorder fully kicks in. In our initial series of 96 patients, the average length of the prodromal period was 22 years.

Once RBD begins, it's generally progressive. Chronic RBD rarely resolves on its own. Instead, the frequency and severity of episodes usually increase over time, whether gradually or rapidly. Most people with the disorder have a major episode at least every two to three weeks, but some have more than one episode every night. Nevertheless, in patients with RBD and a neurodegenerative disorder, such as parkinsonism, the RBD may diminish or even disappear over time as the neurodegenerative process presumably has damaged the brain mechanisms responsible for behavioral activation during REM sleep.

CAUSES OF RBD

Originally, we believed that about half of all cases of RBD were symptomatic, meaning that they were brought on by other conditions, and the other half were idiopathic, meaning that they had no known causes and seemed to occur on their own. We now know that's not quite true.

RBD often develops in conjunction with other disorders. It's frequently seen with narcolepsy (a multifaceted disorder of REM sleep and excessive sleepiness), with both conditions developing in tandem. It's also associated with a host of other neurological disorders, especially Parkinson's disease and parkinsonian disorders, and also subarachnoid hemorrhage, cerebrovascular disease, multiple sclerosis, brain-stem lesions, neoplasms, and head trauma. The common denominator is that brain disorders can damage the REM-atonia circuitry and activate REM-behavior generator centers.

Acute (rapid-onset) RBD may be brought on by medication toxicity, withdrawal from drug or alcohol addiction, drug abuse, acute attacks of multiple sclerosis, or a stroke. Acute RBD generally goes away when the person stops taking the substance or when the medical condition resolves, though it can occasionally develop into chronic RBD.

MYSTERIES OF
CONSCIOUSNESS IN RBD

To lead in to our discussion on the cutting edge of clinical sleep medicine and neuroscience research, we consider this common scenario in RBD: A 52-year-old healthy man who is happily married becomes aware over a period of weeks that his dreams are changing, with an increasing number of unpleasant and very vivid dreams in which a stranger confronts him in a menacing manner or attacks him. Or it could be a pack of wild animals chasing him and attacking him. Then after a period of several more weeks or a few months, this man mysteriously becomes able to carry out these abnormal dreams (he is freed from the "shackles" of REM-atonia) while he is still asleep and in bed with his wife. He sits up in bed, looks around with his eyes closed (attending to the dream action), shouts, throws punches, kicks, or even jumps out of bed while dreaming, and he will hurt himself or his wife from his violent activity. Sometimes in his dream he is fighting to protect his wife from an attacker, but awakens to his wife's frantic screams, and finds his hands grabbing her neck or pulling her hair as he is trying to drag her out of bed. This man then sees his doctor, is referred to a sleep center, is diagnosed with RBD after a sleep-lab study, responds well to therapy (clonazepam)—but in ten years (give or take another eight to ten years) he starts to develop the classic signs of Parkinson's disease, with a persistent hand tremor, muscle stiffness and rigidity, slowed movements, a "shuffling gait" when he walks, and a host of other motor problems.

As we've learned, the first manifestation of parkinsonism can be an alteration of consciousness, specifically the kind of altered consciousness we see in RBD. Then, ultimately, a decade or so later, classic parkinsonism declares itself. In this way, problems with dreaming and sleep behaviors can herald parkinsonism.

Parkinson's Disease, Dementia, and RBD[3]

The strong link between RBD and Parkinson's disease is a major recent discovery in sleep medicine.

For several years, we closely followed the cases of men with presumed idiopathic RBD, and before long it became apparent that these men with

RBD were going on to develop Parkinson's disease at a high rate. In 1996, we analyzed a study that showed that 38 percent of men over the age of 50 with what was originally thought to be idiopathic RBD went on to develop a parkinsonian disorder.[4] Seven years later, the numbers were more staggering: We discovered that 65 percent of this same group of men over the age of 50 who have RBD will go on to develop parkinsonism and/or dementia.[5] Since then, the number has risen to 70 percent, and we're continuing to follow up. No other neurological disorder has emerged, with one exception. Similar findings have been reported by other centers.[6]

The timing is highly variable: In our series of 29 patients, it ranged between 3 and 29 years from the onset of RBD to the onset of parkinsonism and/or dementia. The sleep disorder was the first sign of parkinsonism, which challenges the idea of idiopathic RBD. Some investigators are calling it "cryptogenic RBD." The neurological lesion is there, causing RBD but nothing else, for years until it shows its hand more fully with the classic signs and symptoms of a neurological disorder.

This, of course, is a difficult thing to tell a patient. Not only did the patient and his wife have to cope with a long-standing, dangerous sleep disorder, but now they need to be told about the substantial risk of developing Parkinson's disease, another parkinsonian disorder, or dementia. However, we hope that this piece of the puzzle will lead to advances in understanding parkinsonism. It's possible that early intervention of RBD patients with brain-protecting drugs (being developed) can help further delay the onset or even prevent the emergence of parkinsonism. It is clear that recognition of RBD in the long term will lead to a deeper understanding of the mechanisms responsible for various neurological disorders, which may then lead to an early cure. This is currently an area of very heavy research.

Men over the age of 50 with RBD should have neurological follow-up care on a frequent basis for early detection and treatment of Parkinson's disease and/or dementia.[7]

Parkinson's disease is the most common form of parkinsonism, but there are other disorders that are also classified under the parkinsonism heading—it's a descriptor based on the hallmark symptoms of Parkinson's disease, including resting muscle tremor, *bradykinesia* (slowing of movements), and lead-pipe rigidity. Other types of parkinsonism include multiple system atrophy (MSA), progressive supranuclear palsy (PSP), and dementia with Lewy bodies (DLB).

Multiple System Atrophy and Progressive Supranuclear Palsy

Multiple system atrophy (MSA) is a rare, complex disease with numerous symptoms and a progressive course. It shares traits with Parkinson's disease, including muscle tremor and rigidity, and difficulty walking. However, its damage in the brain is even more widespread and severe, causing additional symptoms, and life expectancy from the onset of the disease is only ten years.

In a 1997 study, researchers in Japan performed overnight PSG studies with video monitoring on 21 patients with MSA, none of whom complained of abnormal sleep behavior. They discovered that all but one patient had an absence of atonia during REM sleep at least 15 percent of the time. In addition, in all but two patients there were motor events during REM sleep, such as facial and limb movements. In another study of MSA and RBD published by an Italian group in 1997, a whopping 90 percent of patients with MSA were documented in the sleep lab to have RBD, leading to the obvious conclusion by the authors that "RBD is the most common sleep disorder associated with MSA."

Progressive supranuclear palsy (PSP) causes problems with balance, control of eye movements, speech, and swallowing, in addition to parkinsonian symptoms. PSP is not inherently life-threatening; however, people are at increased risk of death by injuries from falls. A 2005 study in France of 15 people with PSP, 15 people with Parkinson's disease, and 15 control subjects found that those with PSP had percentages of motor-behavioral abnormalities during REM sleep similar to those with Parkinson's disease.

Dementia with Lewy Bodies

Dementia with Lewy bodies (DLB) has three main hallmarks: parkinsonism symptoms (such as rigidity, tremor, difficulty walking), recurrent visual hallucinations, and periods of confusion and sleepiness alternating with nearly normal thought patterns and alertness. Those with this type of dementia also often have visuospatial impairment, repeated falls, delusions, and transient loss of consciousness. DLB accounts for 15 to 20 percent of all autopsy-confirmed cases of dementia.

A close correlation between RBD and Lewy body disease has been established. In one study, researchers from the Mayo Clinic analyzed 37 people with degenerative dementia and RBD. About half of these patients had signs of parkinsonism and the other half didn't. In 95 percent of patients, RBD either came before or started at the same time as dementia. The researchers discovered that 92 percent of this group had symptoms of DLB. Autopsies on three patients confirmed Lewy body pathology in all three.

Then the Mayo Clinic researchers sought to find out whether the type of dementia seen with RBD was similar to, or different from, the dementia found with Alzheimer's disease (AD). Their study compared 31 patients with mixed RBD and dementia and 31 patients with AD (autopsy-proven) and absence of Lewy bodies, and no history of RBD. The visual vs. verbal memory deficits on "neuropsychological testing" were very different between these two groups: the RBD group had a dementia characteristic of DLB (including visual memory deficits), whereas the AD group had typical findings for AD (featuring verbal memory deficits).

The Mayo Clinic followed this up with a study to determine whether patients who had dementia and RBD but not the other core features of DLB should still be diagnosed with DLB. What if a patient with dementia doesn't have parkinsonism or visual hallucinations but does have RBD? Is it fair to assume that the person has DLB and not Alzheimer's disease or other types of dementia? The researchers concluded, "Results show that early dementia in probable [dementia with Lewy bodies] and dementia with RBD are neuropsychologically indistinguishable." The RBD patients were easily distinguishable from patients with Alzheimer's disease. In a subset of patients, they added, the visual hallucinations and parkinsonism came later—one to six years after the RBD and dementia. Because of this, the researchers, led by Bradley Boeve, M.D., proposed that RBD be added to the list of core diagnostic features of DLB.

The strong association of RBD with parkinsonism makes neuroanatomic and neurophysiologic "sense"—the nerve centers (motor nuclei) responsible for tonic and phasic motor activity and behavioral release during REM sleep are located in the brain stem and connect strongly with the motor nuclei affected by Parkinson's disease. A major question is whether PD destabilizes REM sleep to the point of causing RBD.

Synucleinopathy

The term *synucleinopathy* encompasses all three of the conditions previously mentioned in this section: Parkinson's disease, multiple system atrophy, and dementia with Lewy bodies. They are given this name because of the abnormal presence of densely packed alpha-synuclein proteins that form lesions in brain cells of people with each of the three conditions. It's now suggested that RBD is an important warning sign of synucleinopathy. Those with RBD not attributable to other conditions may be in preclinical stages of the aforementioned diseases.

ATTEMPTED SUICIDE BECAUSE OF RBD

A newly married 35-year-old Taiwanese woman's husband got an unwelcome surprise shortly after the wedding: His new bride often punched him in the face during sleep, usually while she was sleep-talking in an angry manner.[8] The husband moved into his own bedroom and argued with his wife about how her sleep behaviors were disrupting his sleep and putting their marriage in jeopardy. She became depressed and overcome with guilt, and tried to overdose on medications. When she finally saw a sleep specialist and went on an appropriate dose of clonazepam, her RBD remitted and she was able to share a bed with her husband again. Their marital problems ended and her depression lifted immediately.

TREATMENT OF RBD

In nearly 90 percent of cases, clonazepam (Klonopin) is effective in controlling RBD, no matter what conditions may have caused it or occur alongside it. Clonazepam must be used on an ongoing, nightly basis; patients who stop taking the medication usually have relapses of symptoms very quickly, but also quickly get the symptoms back under control when they resume the medication. Both the dream disturbances and the sleep behavioral disturbances of RBD are equally well controlled with clonazepam therapy.

It's not known how clonazepam works to control RBD. When we examine treated patients in PSG studies, we still see that they lack muscle atonia in REM sleep. So their bodies are still *able* to act out their dreams, but they don't, suggesting that clonazepam tones down the "phasic motor overdrive." There may be diminished limb movements and twitching during REM sleep, and the medication generally puts an end to the vigorous and violent behaviors and loud vocalizations that are the most problematic symptoms of RBD.

Clonazepam is effective long term, and to date no major complications have arisen from its use. Patients who've been on it for many years to treat RBD have shown remarkably minimal dosage tolerance, meaning that they don't usually have to increase their dose over the years despite the body getting "used to" the medication.

In our clinic, we give patients information that allows them to make adjustments in their dosage, within a safe range, in their effort to find out the optimal dose and the best time of night to take it. They may take clonazepam anywhere from one hour to ten minutes before bedtime, depending on what works best and minimizes oversedation in the morning. Some patients have been able to keep RBD under control by taking the medication only every other night, but this is uncommon. Usually, patients take 0.5 mg of clonazepam to start, then increase the dose up to 0.75 or 1 mg, as needed. Some people take 2 mg, and even up to 4 mg, if it's not fully effective at the lowest dose. We encourage them to split pills and find the ideal dose for themselves.

For those who don't have success with clonazepam alone, melatonin may help. Preliminary research indicates that perhaps it helps restore the normal atonia of REM sleep. Dopamine-enhancing medications, such as pramipexole, may also be of benefit. There is a long list of tertiary medications for treating RBD, but these rarely need to be used.

THE FUTURE OF RBD

Twenty years ago, our clinic was the only one talking about RBD. It took a few years for other doctors and sleep researchers to take us seriously, but now RBD is well acknowledged as a major disorder in sleep medicine, situated at a very busy crossroads of the neurosciences and sleep medicine.

There are so many questions yet to be answered (and also yet to be raised) with RBD. For instance, we now wonder whether the visual hallucinations present with parkinsonian disorders and dementia with Lewy bodies are actually intrusions of the dream imagery of REM sleep into wakefulness. And might sleep attacks in these disorders be REM sleep disturbances? If so, would narcolepsy treatments benefit people with parkinsonism or dementia?

If RBD is a harbinger of synucleinopathy, how could it help us solve the puzzle of why the alpha-synuclein proteins "go bad" and attack some people's brain cells? Why does RBD emerge one or two decades before classic parkinsonism in some patients but emerges concurrently with parkinsonism in other patients, or even emerges after classic parkinsonism in yet other patients?

Because RBD may lead to much earlier detection of parkinsonian disorders and dementia, we now need to figure out how that can help us: Is there any form of early treatment that may reverse or slow the progress of these conditions? We hope that the discovery of RBD is a stepping stone in the future of neurological science, and will lead to improved treatment of parkinsonism and dementia.

A current pressing question is whether medication-induced RBD—as with the Prozac-like SSRI drugs, and other medications (such as Effexor [venlafaxine], Remeron [mirtazapine], and tricyclic antidepressants)—carries any risk for future parkinsonism. The same question about increased risk for future parkinsonism applies to the *parasomnia overlap disorder* (POD) in which a person is afflicted not only with RBD, but also sleep-walking and/or sleep terrors.[9] POD can be associated with a variety of medical disorders (including neurological and psychiatric) and medications. It is interesting to note that Wellbutrin (bupropion), which is a dopamine-enhancing antidepressant, has not been reported to cause or aggravate RBD. Serotonin-enhancing antidepressants, on the other hand, can disturb REM atonia and increase the risk for RBD.[10] Finally, are women who are afflicted with RBD also at risk for developing future parkinsonism?

Chapter 14

NOCTURNAL DISSOCIATIVE DISORDERS

The Lost Children as Adults

At this point in the book, if I asked you to imagine a woman wandering outdoors in her nightgown in the middle of the night and trying to hail a taxi, what would your presumed diagnosis be? Sleepwalking, right?

Some people—typically females who've experienced severe childhood trauma, including physical, sexual, and verbal-emotional abuse—have episodes of behavior after they've gone to bed at night that they don't remember the next day. Although many people with parasomnias don't recall what they did during the night while asleep, people with nocturnal dissociative disorders (DDs) have amnesia for what they did in the midst of "dissociated" states of wakefulness during the night—"islands of wakefulness" surrounded by sleep that may represent the partial or elaborate reenactment of past abuse scenarios, including the urge to escape. In other words, with a nocturnal DD, you don't remember what happened when you were awake, but with parasomnias, you don't remember what happened when you were asleep or immediately after you were abruptly awakened.

To understand nocturnal dissociative disorders, we must first understand the concept of *daytime* dissociative disorders.

DISSOCIATIVE DISORDERS DEFINED

To *dissociate* is to have your wakeful conscious self be split off from a situation. The classic scenario in the development of a dissociative disorder involves a child who is subject to repeated abuse by a parent or caretaker: After a while, the child automatically dissociates to cope with the abuse. He or she may "block out" the experience and daydream of being somewhere else or someone else, or hear birds chirping or cats meowing rather than hearing the horrible predatory sounds of abuse. When the abuse is too terrible to deal with, the child may unconsciously make it seem unreal, utilizing the psychological defense mechanism of dissociation to escape the situation for a while.

When the terrible experience is over, the child doesn't have to incorporate it into her or his conscious memory—it may be successfully blocked out (*repressed*), or it may feel like it happened to someone else or just wasn't quite real. I should point out that the capacity to dissociate is something we seem to be born with, and not a coping strategy that we consciously can generate. Nevertheless, if we do have the inherent capacity for dissociation, then we unconsciously utilize it under stressful circumstances—to create a distance between ourselves and an abuse situation, or the memory of abuse, which serves the purpose of lessening the intense anxiety and angst produced by the abuse. Therefore, dissociation can serve an adaptive or useful purpose, even though too much of it is not good, since then too much of our lives is riddled by disconcerting memory gaps, with consciousness being subject to fits and starts. This is not good for keeping your job or staying in school, and it is not good for your social life.

Abuse isn't the only type of trauma that can lead to a dissociative disorder (DD). Natural disasters, deaths, medical ordeals, and other frightening and distressing situations can trigger these disorders. Normally, though, a one-time event won't do it: The trauma must be repetitive or ongoing for a person to use the psychological defense mechanism of dissociation on a regular basis, and have it become available as a coping mechanism.

Because of its strong association with trauma, it's not surprising that most people with dissociative disorders also have diagnoses of post-traumatic stress disorder (PTSD). A core feature of PTSD is *emotional numbing,* which sits at one end of the spectrum of symptoms. On the opposite end is *emotional flooding* (when traumatic memories are intrusive). If a person experiences emotional numbing too much, he or she is vulnerable for the next step, which is dissociation, with recurrent and prolonged states of emotional numbing.

Once a person has developed a dissociative disorder, the dissociation no longer has to be provoked by a traumatic event—the person may dissociate at any time of stress or even when seemingly unprovoked. In other words, dissociation can take on a life of its own—like a brain-mind software program turned on so many times that ultimately it can turn itself on—and can greatly interfere with one's personal life and promote maladaptive, self-defeating behaviors. For example, if one keeps forgetting about appointments or other planned activities, then there will be negative repercussions in the occupational and social spheres of one's life, which can then result in demoralization and a deepening of depression and confusion.

Also, from a clinical perspective, whenever a patient reports having few or no memories about a certain time period lasting weeks, months, or longer in their childhood or some other time in their life, that is a red flag for probable abuse, dissociation, and memory suppression. In fact, clinicians (physicians, nurses, psychologists, social workers) should more frequently ask about such memory gaps during their intake interviews or during follow-up sessions. People with significant memory gaps for periods of their life often suffer from long-standing, low- to medium-grade depression, called *dysthymia*.

WHAT HAPPENS AT NIGHT

Just like the example I gave in the beginning of this chapter, patients with sleep-related DDs usually come to the clinic because their doctors believe they are sleepwalkers, based on the observation by others who have witnessed their unusual and often elaborate episodes.

The woman at the beginning of the book is based on a real case we saw at the clinic. After a comprehensive clinical evaluation including an overnight PSG, psychological testing, and psychiatric interviews, we found that the woman actually had nocturnal DD, which looks much different on sleep tests than sleepwalking. At night, her behaviors occurred during a wakeful dissociated state that happened after she'd slept for some time. She also had an early life history of multiple forms of abuse that often were inflicted during the evenings or in the night.

In 1988, we presented this woman's case at the Associated Professional Sleep Societies' Fourth Annual Meeting, and a year later, we detailed eight cases of sleep-related DDs—the first series of cases to be reported in the medical literature.

Of the eight patients we profiled, one had DD that appeared only during sleep hours, and the rest also had DD during the daytime (an around-the-clock DD).

"Cat Boy"

The patient with exclusively nocturnal DD was also the only male we studied. He was affectionately referred to as the cat boy. He was 19 and had been having the same strange spells once or twice a week at night for four years. An hour or two after he went to sleep, he'd get out of bed and "become" a large jungle cat.

When I say "become," I mean that he adopted all the behaviors and physical attributes of a jungle cat that a human could possibly do—it was impressive how convincing he was, as was evident on a home video that his parents brought in to our clinic. They taped it to prove that they were not exaggerating about what their son did. The same behaviors were recorded on audiovisual tape in the sleep laboratory. No doubt about it: He was fully transformed into another species and was on the prowl at home and in the sleep lab as a large jungle cat. Even his hands were stretched and bent into surprisingly realistic-looking paws.

He would growl, hiss, crawl, and leap about for up to an hour at a time. And he'd have feats of "superhuman strength," as his family explained: He could bite down on a mattress and drag it across the room with his teeth, or lift a marble table with his jaws. He would never stand up during these episodes; everything he did was on his hands and knees, and he would use his mouth to pick things up and move objects.

On questioning, he told us that he was having the same "dream" every time these episodes occurred: He was a lion or tiger, and a female zookeeper had let him out of his cage. She held a piece of raw meat, and he followed her down a path, trying to snatch the meat. But there was an "invisible force field" that prevented his ability to grab the meat, which frustrated him. The dream always ended the same way: Someone would shoot a tranquilizer gun at him, and he'd fall down and become unconscious. The only variable in the dream was the zookeeper's face; she looked different each time, though never a familiar face. During these dreams, he told us, he felt completely like an animal, and had no memory of being human.

The "dream" he described was much the same as his real actions. He was typically unresponsive to his family, even if they splashed cold water on his face, yelled at him, or shook him.

Soon after his sleep problem began, he had been in a psychiatric unit for a month for evaluation and treatment. Doctors decided he was having a major depressive episode, but various antidepressant medications and therapy didn't have any effect on his sleep behaviors. Long-term outpatient family therapy and individual psychotherapy didn't help, either. Then doctors tried him on two different psychotropic drugs, tranylcypromine and clonazepam, both of which paradoxically aggravated his sleep-related problem instead of improving it.

His parents and sister came with him to the sleep lab, and showed us the videotape they'd made at home of one of his episodes. The young man hadn't seen the tape yet . . . and he began crying in disbelief when he saw his behavior. So we did two consecutive overnight PSG studies to find out what we could about his condition.

On the first night, he had two "jungle cat" occurrences. The first was just under an hour after he fell asleep. For two minutes, he lay in bed and growled intermittently, then he abruptly got out of bed and crawled around the room while growling and hissing and grinding his teeth. He pulled on the mattress with his jaws and chewed and swallowed part of the airflow-monitoring device. After 6.5 minutes of this behavior, he collapsed on the floor for 2 minutes, then began crawling and growling again for another 4.5 minutes. Again, he collapsed, and was unresponsive for 30 seconds. Then he became alert and described what he had been "dreaming" in full detail, though he didn't remember acting anything out.

One hour and 15 minutes later, he was up and at it again—growling, crawling around the floor, knocking over a lamp, swallowing part of another airflow monitoring device, and repeatedly banging his head against a wall. After more than nine minutes, he collapsed on the floor just like he had earlier, and after 20 minutes, he reported that he had been having "the same dream again, nothing different."

The EEG showed wakefulness or drowsiness and no seizure activity throughout all the times he was acting like a jungle cat. In other words, his brain was awake, but it was dissociating.

This young man was adopted when he was 10 months old, and although there was no abuse (but rather much love and nurturing) after adoption, his prior history is unknown, and so very early life abuse/privations remain a possibility. Although the DD diagnosis raises suspicions of early life abuse, another possibility involves the derepression of *atavistic genes,* ancestral phylogenetic genes that we all carry. Mysteriously, these genes may become activated in a vulnerable person and can come out shortly after sleep is interrupted, instigating animalistic behaviors—taking charge of our bodies and minds and making us act like jungle cats or wolves or monkeys. But another indicator that this case may have been brought on by early trauma is the fact that even at age 19, this person had severe separation-anxiety attacks whenever his mother was away from home at night. His sister told us that he showed regressive behavior when their mother slept away from home.

We offered treatment in the form of hypnotherapy, in which he would learn to visualize in a hypnotic state as he was falling asleep that he would remain in bed and sleep through the night without getting up. The hypnosis would not involve thinking about his animal behavior and the need to suppress it. All he would need to concentrate on were the positive expectations of remaining in bed all night. One parent, however, was overly skeptical and hypersensitive about this idea, and considered it another form of "psychological therapy," which previously had not cut back on these episodes.

I have had two brief follow-ups with his physician since that time, and the episodes had become progressively less frequent and less elaborate, without any specific therapy. His physician informed me a few years ago, about 15 years after I had evaluated him, that his parasomnia eventually completely disappeared. Perhaps the animalistic genes became repressed again. What a strange, fascinating, and truly awesome case. A radical day/night split personality.

Extension of Daytime DD

The other seven cases in our original report were women who had daytime dissociative disorders along with their nightly episodes. One, a 34-year-old woman, was referred to us by two different psychologists, each of whom had no idea she was seeing the other. This woman had a nearly unbelievable history of trauma, as verified by her sister.

The woman's first husband died in an accident, and another sister died in a car accident. Then, within a nine-day period, her father died of myocardial infarction, her mother committed suicide, she had a miscarriage, and her second husband left her. In addition to all that, she was hospitalized for a heart condition, and had various other health conditions.

Five months before she was referred to us, she began having sleep-related episodes at least twice a week. During most of the episodes, she would wander around her backyard or the streets while wearing a nightgown. Often, she'd injure herself by running through a glass door or crashing into a window, or running into furniture. Sometimes, she'd do other odd activities like cutting her hair in a bizarre fashion, shredding her daughter's new snowsuit, or tearing down drapes. But twice, the "odd" behavior became truly terrifying.

Both times, she managed to drive to an airport, purchase a ticket to a distant city, and get on a plane. She was dressed but had no luggage. Both times, she "awakened" during the flight and immediately had a panic attack—where was she? Where was her two-year-old daughter? As soon as the plane landed, she got on a return flight and went back home. She had complete amnesia for how she got on a plane, or even out of the house.

By the time she came to see me, she had hired a live-in housekeeper to protect her daughter from any dangerous spells and watch her during any nocturnal wanderings. She had also been on a variety of medications for post-traumatic stress disorder, depression, and panic disorder; and had been on clonazepam and diazepam, none of which helped her sleep-related spells.

She had begun having dissociative symptoms during the day, including a tendency to wander to the maternity ward of a local hospital and stare at the babies through the window. Although she said she didn't recall doing this, she did say that she was searching for her latest miscarried baby, who she was sure was still alive.

Because of her dissociative symptoms, we suspected she might have a nocturnal dissociative disorder, but this had to be confirmed—and it was. During an overnight PSG study, we didn't see any seizure activity or REM abnormalities, nor any abnormal arousals during NREM sleep. But we did see two episodes during clear-cut EEG wakefulness after she had awakened: Once, she walked down the hall and spoke in a confused manner to the sleep technician, and another time, she sat up in bed, hyperventilating, and wasn't able to talk for three minutes before becoming coherent and then going back to sleep.

Somehow, this woman held down a high-level job in the computer industry even as her disorder grew worse and worse. Seven months after the disorder reared its head, her sister told us that the woman had just begun referring to herself by different names, and assuming at least three different personalities: childlike, conservative/highly competent, and rageful. Each alter ego had its own voice and mannerisms. Soon thereafter, her sister found an "unusual assortment of clothes" in the patient's closet, ranging from conservative to "completely atypical, glow-in-the-dark, horrible pink dresses." She also found credit cards in three different names, but with the same billing address.

The woman was diagnosed with *dissociative identity disorder* (formerly known as *multiple personality disorder*) and *psychogenic fugue* (which is dissociation that leads the person to take trips away from home—in this case, the two airplane rides). We were not successful in treating her. She had a panic attack during hypnotherapy and abruptly walked out of the office, and refused to engage in long-term psychotherapy at our hospital because she was convinced that the therapist would die like her father did.

Several months later her daughter died of a heart condition at age three, and the woman's nocturnal wanderings became more frequent. She would wander into the yard and stare blankly several times a week. On last check, she had moved to a new city, nearer to her sister.

The remaining women with nocturnal dissociation, all in their twenties and thirties, had less extreme episodes, often involving wandering and injuries and sometimes violence to the bed partner or self. Some of the episodes were clear reenactments of abuse scenarios, with groaning noises and defensive body movements and associated "dreams" about being abused. Sometimes these would be sexual behaviors, sometimes physically violent, and would be paired with statements such as, "Don't do that; you're hurting me. Get away." Others' episodes consisted of agitated or confused behaviors, though not overtly abuse-related.

When we see people with sleep-related dissociative disorders, they sometimes believe they're acting out their dreams . . . but in fact, they're reenacting painful, dissociated wakeful memories. Perhaps it's easier on the psyche to believe that the thoughts and bad memories are dreams, and safer to think about them in that context, rather than recognizing that they're occurring in wakeful reality.

In the case of dissociative identity disorder, sleep may be a conduit to changing personalities. The person may fall asleep as one "personality" and wake up in the middle of the night as another, having no memory of this the following day. Since sexual abuse often occurs at night in bed, it is not surprising that dissociative disorders can emerge in relation to nocturnal sleep. The symptoms can span the range of reexperiencing a headache or panic attack from past abuse to a full reenactment of the abuse.[1, 2, 3, 4]

WHAT ELSE COULD IT BE?

There are several differential diagnoses for nocturnal dissociative disorders, including nocturnal seizures, night terrors, sleepwalking, RBD, and dream-interruption insomnia—the latter of which involves repeated awakenings from most REM periods, with frequent recall of disturbing dreams. As with other sleep disorders, malingering should be considered. A PSG study is important in the diagnostic process.

Dissociation by the Numbers

In 2001, a group of researchers at a university in Turkey set out to determine the frequency of nocturnal dissociative disorder in patients with daytime dissociative disorders, and to compare the clinical characteristics of those who did have nocturnal episodes with those who didn't.

They evaluated 29 consecutive patients with DD and found that eight of them also had nocturnal dissociative episodes, making the prevalence 27.5 percent.[5]

Some of the characteristics between the two groups were very similar. The average age of onset of dissociative disorders was 19.6 in the group with nocturnal episodes, and 18.9 in the group without. About 38 percent of both groups had attempted to commit suicide, and approximately equal percentages of both groups had major depression and borderline personality disorder.

Yet other characteristics differed. The most obvious was violence during the sleep period: One hundred percent of those with nocturnal dissociative episodes had violent behavior during sleep, compared to zero percent of those without nocturnal dissociative episodes. Half of the first group also had self-mutilating behavior during the sleep period, compared to none of the second group. Nearly all—87.5 percent—of the first group also experienced sleep-related hallucinations, compared to 48 percent of the second group.

Their trauma histories differed a bit, too. Half of those with nocturnal episodes experienced physical abuse, compared to 38 percent of those without nocturnal episodes, and 62 percent of the former group experienced sexual abuse, compared to 76 percent in the latter group.

SLEEP-RELATED DISSOCIATIVE DISORDER IN SUMMARY

The researchers in Turkey proposed the term *pseudo-parasomnia* to describe nocturnal dissociative disorder, because true parasomnias arise suddenly during arousals from sleep or else within sleep, whereas an episode of nocturnal DD arises from well-established wakefulness (a minute or two after an awakening).

However, in our center's series of cases, even though we can see from the EEG that the person is awake, there may be no immediate behavioral change. The person remains lying still in bed, just as she or he had been lying for the previous 12 to 15 minutes. So even though there's been a major state change, oddly, the person doesn't seem to notice or respond to it in the way that most people would—if you wake up in the middle of the night, you'd probably open your eyes, roll over, stretch, or even get out of bed. But the person with DD still appears to be asleep. Within a matter of minutes, though, the dissociative "brain software" kicks in and the movements begin—with a very particular behavioral repertoire during the nocturnal DD episode.

In our study of eight patients, seven of the eight had PTSD stemming from early childhood trauma.

It's not clear why some people with dissociative disorders experience them during the sleep period and others don't. It may be that those with sleep-related dissociative episodes have trauma experiences that are more linked with the bedroom or sleep, along with a biological vulnerability for wakeful dissociation during the sleep cycle.[6]

But again, it's important to note that sleep-related dissociative disorder is *not* sleepwalking, and shouldn't be referred to in that manner. It's a separate disorder that occurs when people are clearly awake, and which needs specialized psychiatric therapy. It is a true "psychiatric parasomnia."

Finally, though a PSG is strongly recommended, in only about half the cases will the findings be conclusive in capturing a DD episode during the night, either arising during wakefulness in the immediate pre-sleep period or else during an awakening from sleep later in the night. A PSG study, however, can help exclude other diagnostic possibilities, or positively identify another cause of the problematic nocturnal behavior. Clinical interviews, psychological testing, and other "detective work" remain the cornerstone for diagnosing a nocturnal DD and for pinpointing other causes of troublesome nocturnal behaviors.

Chapter 15

SLEEP-TALKING

Disruptive Chatter, Obscene Shouting, and Desdemona Scenarios

S leep-talking, or *somniloquy*, can occur in any stage of sleep. Some people carry on complicated monologues or dialogues with their bed partners in their sleep, while others speak gibberish or just mumble a few words. It can occur on its own, or in conjunction with other sleep disorders—particularly sleepwalking, sleep terrors, and REM sleep behavior disorder. Sleep-talkers are not critically aware of their behaviors, or how they might sound to others. Their voices and the type of language they use may sound very different from their wakeful speech.

It's common in adults and even more common in children—in fact, 50 percent of young children have sleep-talking behaviors. A survey of 2,022 schoolchildren from ages 3 to 10 concluded that about half of children in that age range sleep-talk at least once a year, but less than 10 percent do it every night.[1]

Rarely, adult-onset frequent sleep-talking is associated with a psychiatric disorder, but in most cases, there is no serious psychopathology. Also rarely, sleep-talking that is *stereotypic,* or repeated just the same way every night, may be a result of nocturnal seizures. Sleep-talking and shouting also occur with various other parasomnias, such as RBD, sleepwalking, sleep terrors, and confusional arousals with sleepsex talking.

SLEEP-TALKING IN
RELATIONSHIPS

I encounter a few young married couples every year with the same dilemma: Shortly after they get married, one or the other sleep-talks about a former boyfriend or girlfriend. The spouse wonders, "What does it mean?" Does it mean he or she is still in love with that other person? The honest answer is that we don't know, but we can't jump to conclusions. People sleep-talk about all sorts of things; some of the talking makes no sense at all, and some of it may relate to past events, experiences, and relationships that no longer have any current relevance or emotional impact—it's just spilling out from the memory banks without discrimination or censorship. This is why the analysis of sleep-talking content often has dubious validity—the uncertainty factor is too great.

It's the same as with dream interpretation: Yes, some dreams are meaningful, but others are just off-the-wall. As much as we'd like to find logic and significance in the abstract meanderings of our brains in altered states of consciousness, you can't jump to the conclusion that there's any emotional involvement with a previous boyfriend or girlfriend just because of sleep-talking or dream content. Is there any evidence that your partner is not fully engaged in your relationship or dwelling on a past relationship? That's much more important than sleep-talking.

Some people tell strange lies in their sleep. One woman reports that she would often tell tales of her sexual exploits while sleeping. She'd give detailed accounts of her affairs, and her partner was very upset by this. But she never had any of these affairs, and had a hard time convincing her partner that the sleep-talking was nonsense.

Modern sleep science and the law accept that sleep-talking is not a product of a conscious or rational mind, since the person is asleep and therefore unconscious, although the extent from person to person, and during the various sleep stages, has great variance, which adds to the intrigue about how parasomnias occur.

Desdemona's Tragedy

Shakespeare's classic play *Othello* gives a perfect example of how much weight a person can give to sleep-talking as evidence of betrayal—and how wrong that judgment may be.

In the play, Othello is an army general. Iago is a soldier who has just been passed over for a promotion, and vows vengeance. Othello has chosen Cassio for the lieutenant position, and now Iago is ready to make both of their lives miserable. He sets out to make it appear that Othello's wife, Desdemona, is having an affair with Cassio. First, he enlists a sidekick to get Cassio drunk on duty and start a fight. When Othello sees this, he angrily dismisses Cassio. Iago counsels Cassio that the only way he'll gain Othello's favor again is if he gets Desdemona to plead for him—which he does.

Iago quickly seizes on this opportunity to raise doubts in Othello's mind: Why was Cassio sneaking around to talk to Desdemona? Why is she pleading for this man? Might she be in love with him? Othello's suspicions are raised, but he soon decides he should trust his wife and won't believe unless there's evidence of cheating.

Othello
Villain, be sure thou prove my love a whore,
Be sure of it; give me the ocular proof:
Or by the worth of man's eternal soul,
Thou hadst been better have been born a dog
Than answer my waked wrath!

Iago provides a simple answer: He caught Cassio sleep-talking about the affair.

Othello
Give me a living reason she's disloyal.

Iago
I do not like the office:
But, sith I am enter'd in this cause so far,

Prick'd to't by foolish honesty and love,
I will go on. I lay with Cassio lately;
And, being troubled with a raging tooth,
I could not sleep.
There are a kind of men so loose of soul,
That in their sleeps will mutter their affairs:
One of this kind is Cassio:
In sleep I heard him say "Sweet Desdemona,
Let us be wary, let us hide our loves";
And then, sir, would he gripe and wring my hand,
Cry "O sweet creature!" and then kiss me hard,
As if he pluck'd up kisses by the roots
That grew upon my lips: then laid his leg
Over my thigh, and sigh'd, and kiss'd; and then
Cried "Cursed fate that gave thee to the Moor!"

Othello
O monstrous! monstrous!

While only a minute earlier he had demanded "ocular proof," now he relies only on the secondhand account of a sleep-talking incident made up by a vengeful source. This was enough proof needed for Othello to declare, "I'll tear her all to pieces."

And he does: He suffocates her with a pillow, and orders Cassio to be killed, too. Only after Desdemona is dead and Cassio wounded does Iago's plot come to light. Overcome with guilt, Othello then kills himself with his dagger, with one last kiss for the loyal wife he just murdered.

There is a broad spectrum of possible adverse psychosocial consequences from sleep-talking, though the Desdemona scenario is obviously the most extreme outcome.

DIVORCE! DIVORCE! DIVORCE!

In 2006, a Muslim couple in India were told they had to divorce because of the husband's sleep-talking.

Sohela Ansari confided to her friends that her husband, Aftab, had said the word *talaq* in his sleep three times.[2] *Talaq* means "divorce." When local religious leaders heard about it, they told the couple that Aftab had divorced his wife in his sleep, due to a custom known as "triple *talaq*." The Muslim clerics issued a decree telling the couple they were to separate immediately. Despite the fact that they were married for 11 years, had three children, and did not want to divorce, the leaders said they had to—and that if they wanted to remarry, they'd have to wait 103 days and Sohela would have to first marry another man and spend a night with him before the new husband divorced her. Then Sohela and Aftab could remarry, a judge ruled.

Although they were legally divorced, they refused to obey, and visited a family counseling center at a police station for advice. Islamic scholars disagree about the interpretation of the law in this case—whether it should be upheld, considering the man was in a sleep state.

SLEEP-TALKING AS EVIDENCE

Judges have been called upon to decide whether sleep-talking is admissible evidence in courtrooms. In one publicized case, Jorge Almeida was convicted of sexual molestation in part because of a man's testimony about what the alleged victim, who was his daughter's friend, said in her sleep at a sleepover: "Jorge, get off me. Jorge, get off me."[3]

While this was not the only evidence in the case—the girl took the stand to testify about the alleged attack—the conviction was overturned by the Supreme Judicial Court because they determined that the sleep-talking should not have been allowed as evidence since there's no way to prove its reliability. In another case, a mother testified that she first learned her son had been sexually assaulted by his step-grandfather because he talked about it in his sleep.[4] The mother said that the boy frequently talked in his sleep about the day's events. In this case, he was begging his step-grandfather not to pull down his pants and saying that they were supposed to be looking for a Christmas tree—which is just what he was doing with his step-grandfather that day. She asked the boy about his sleep-talking, and he told his mother what had happened in detail.

The man was convicted of aggravated sodomy, but appealed it on the grounds that the sleep-talking was not reliable evidence. The appellate court upheld the verdict.

One of the reasons they upheld this verdict was the boy's usual predisposition to talk about the true events of his day in his sleep. Note that this varies greatly; some people do sleep-talk about their real experiences, while others talk about things not remotely connected to reality.

Shelagh's Story

My husband claims not to dream, and never remembers any wisps or disjointed scenes. But he most certainly dreams, because when we were fairly newly married, sleeping nude one summer night, he cupped my breast and said, "Nice."

Hey, we were young. Middle of the night was fine by me.

I was taken aback when he added, "Got another one of these? I ate the first one up already."

In the morning, of course, he had no recollection of what he'd said or of any dreams.

I will forever wonder what he was dreaming.

THE SLEEP-TALKING COUPLE

One of the funniest confusional-arousal experiences I've heard about—though it wasn't funny to them—came from a married couple, both CPAs and type A personalities. One had a history of sleepwalking and confusional arousals, and the other had no notable history. Well, one night the wife sat up with a confusional arousal and began sleep-talking. This triggered a confusional arousal in her husband, who also sat up and began talking. The two of them were bewildered and feeding off each other, having a strange conversation that escalated in complexity—and eventually they began shouting at each other. They lived in an apartment complex, and the neighbors woke up and were about to call the police. A neighbor came over and banged on the couple's door, which caused one of them to snap out of the confusional arousal and answer the door.

Both of them were horrified—they had no idea they'd been talking, much less shouting, in the middle of the night. And they were very thankful that the police hadn't yet been called. How would they explain this? That's what brought them to see me: They were very worried that the next episode would result in a police visit, and they were embarrassed that the neighbors had heard them.

> The man was alone, in bed. A lamp was burning dimly. He was asleep and talking in his sleep.
>
> He smiled and said "No!" three times with growing ecstasy. Then his smile at the vision he saw faded away. For a moment his face remained set, as if he were waiting, then he looked terrified, and his mouth opened. "Anna! Ah, ah!—Ah, ah!" he cried through gaping lips. At this he awoke and rolled his eyes. He sighed and quieted down. He sat up in bed, still struck and terrified by what had passed through his mind a few seconds before.
>
> —From *The Inferno* by Henri Barbusse,
> translated by Edward J. O'Brien (1918)

STATE-DEPENDENT LANGUAGE ACTIVATION

One of my first patients with RBD, Mel Abel, was a sweet, gentle, and funny man who was universally liked. Just before he went to sleep every night, he'd say his nightly prayers. As he would say, "I often fell asleep with a prayer on my lips." Yet just after 4 in the morning, this happy man would unleash a torrent of invectives in his sleep.

My documentary, *Sleep Runners,* plays sleep-lab footage in which Mel can be heard shouting this strange monologue while being in REM sleep: "Quit using the goddamn bowl for banging like that. Quit it now! Get the hell out of here! Go on! That's about four times this morning that I have told you. I don't know if you're that deaf or that dumb, which. Goddamn continuously . . . What the hell are you looking for, a walleye?"

For Mel, the REM sleep state brought out a loud, foul mouth that did not come out while he was awake, with his wife, Harriet, being the astonished witness. So Mel exhibited in an offensive manner how completely different his brain "software programs" for verbal expression during REM sleep were in contrast to wakefulness. The tone of voice, volume, rhythm, in-flections, content, and duration of speech were radically distinct. This is a common finding with RBD. About 25 percent of men who eventually develop RBD have histories of progressively loud and prolonged (and at times profane) sleep-talking over the course of decades before they start engaging in aggressive and violent dream-enacting behaviors.

A strange and fascinating twist to this phenomenon was illustrated in a 1996 report by Dr. Juan Pareja and his colleagues in Madrid.[5] A 70-year-old woman went to her doctor with the contrasting dual complaints of pronounced sleep-talking and daily problems with inhibition of her speech during her waking life! Imagine: sleep-talking effortlessly in the night but not being able to talk during the day. Once again, we see that different "software programs" can run the show during different states of sleep and wakefulness.

As it turns out, this woman had a neurodegenerative brain disorder that was producing these contrasting effects on her speech production. Her sleep-talking, along with excessive muscle twitching, occurred during REM sleep, so she was diagnosed with early RBD.

SLEEP-TALKING AND MULTIPLE SYSTEM ATROPHY

Sleep-talking during REM sleep was documented during PSG monitoring in 86 percent of a series of 21 patients with multiple system atrophy (MSA), a "parkinsonism plus disorder." Sleep-talking began or intensified at the time of clinical onset of MSA among these 18 patients.[6]

Sleep-talking Hits Home

Just before I began this chapter, I was again reminded of how easy it is for parasomnia behavior to creep into our lives—even for those of us who don't have sleep disorders. I gave a lecture at the Sacred Heart Hospital in Eau Claire, Wisconsin, in January 2006. Afterward, my wife, Andrea, and I went out to dinner with friends, and we each had two glasses of wine with our excellent Italian dinner. Later, at the hotel, we spent about 45 minutes talking in bed about the book that Andrea had just finished reading and the book she had just started to read. Besides listening to her descriptions and comments, I gave my impressions and actively participated in the conversation.

I eventually dropped off to sleep quite rapidly (within a minute or two of my last saying something to Andrea). What she then observed almost immediately after I fell asleep was a threefold shift: 1) I started moving around in bed quite a bit, whereas before falling asleep I was lying still while talking with her and would periodically shift my body position. 2) I began sleep-talking, in which I was audible and spoke intelligibly in a series of short phrases—but the content and tone had nothing to do with all that we had been discussing for the previous 45 minutes or so. I would say short phrases like "Over there" or "There, there" or "Nice job" or "Can't you find it?" 3) There was a major shift in my mood; I began to laugh while talking, whereas during a rather sober conversation before falling asleep about two dead-serious books, my expression and tone were the opposite of lighthearted and there was no laughter.

Andrea thought that somehow she had disturbed me from my sleep and that I was awake when I was talking to her, but she quickly realized that I was sleep-talking while moving around in bed. In fact, I had fallen asleep on top of the covers, and she had a hard time tucking me under the covers, because I was fast asleep. She found all of this quite unusual and told me about it the next morning.

I had no memory of falling asleep, nor any memory of sleep-talking, laughing, or dreaming at any point of the night. So there was a complete disconnect between what I was saying before I fell asleep and what I started saying within a minute after falling asleep—another example of state-dependent language activation. Moreover, there was state-dependent mood activation and also state-dependent behavioral-motor activation. And so a curious little sleep situation emanated from me early on that night—several hours after I had lectured on sleep disorders!

Bilingual Sleep-talking

In 1988, a research team in Spain led by Juan A. Pareja set out to determine what happens when bilingual people sleep-talk—do they always talk in their native language?

They surveyed parents of 681 children in the Basque region of Spain, where about 25 percent of the population is bilingual, speaking Spanish and Euskera, two completely different languages.[7] To be eligible for the survey, the child's teachers and parents had to agree that the child was fluent in both languages. The researchers defined the native language as the one learned first in the home, which, in general, was also the one the children used predominantly with family and friends. If the child learned both languages simultaneously and demonstrated equal proficiency in both, then the child was termed a "balanced bilingual," whereas those with greater proficiency in the native language were termed "dominant bilinguals." Most of the children in the study fell into the dominant-bilingual category.

About half of the children in the study were prone to sleep-talking, and as the researchers predicted, most of the dominant bilinguals spoke their native language when sleep-talking: the dominant language of wakefulness remained the dominant language in the sleep state. The balanced bilinguals tended to speak both languages without preference during sleep. However, a small group of dominant bilinguals (about 4 percent) spoke the nonnative language regularly during sleep-talking episodes. Why?

There's no definitive answer, but the researchers explain several interesting theories. First, it may be that the child is addressing a nonnative speaker in sleep. But there are slightly more complex theories, too.

The researchers point to previous studies showing that bilingual people with auditory hallucinations often hear the "good" voices as speaking their native language and the "bad" voices speaking the secondary language.[8] They suggest that, because emotional stress may set off sleep-talking, the children may express stressful experiences in their nonnative languages.

The sleep-talking may also be set off by dreams, which can occur in either language, depending on the dream content. In the case of RBD, we know that there's often a state-dependent language shift—such as with Mel Abel, the kindly gentleman who would unleash angry monologues during REM sleep after he'd said his nightly prayers—so it's possible that RBD could set off a language-dominance shift as well. If children experience stress more in the nonnative language, it makes sense that they'd use that language in response to scary or aggressive dreams.

Finally, sleep-related dissociative disorder could play a role. As I mentioned, it's been reported that people with dissociative identity disorder sometimes switch alter egos in the sleep state (awakening in a different alter ego), so it's possible that the alter egos have different dominant languages.

Sleep-talking in Summary

It's probably a waste of time trying to determine meaning in sleep-talk. Most of what's said is nonsense, and the "incriminating" things that may be said during sleep are generally not indicative of what's actually on the person's conscious mind. During lighter sleep stages, the talk is more likely to be intelligible, even if it's illogical. In general, no treatment is necessary.

While it's not physically harmful, it can certainly cause embarrassment and can annoy a bed partner or roommate, or be disruptive in group-sleeping situations (such as college dormitories, military barracks, fire stations, or a tent while camping). Because of this, sleep-talkers are sometimes afraid to sleep away from home. Sleep-talking can also cause insomnia in a person sleeping nearby.

Although there's no cure, sleep-talking is less likely if you're not sleep-deprived, haven't had alcohol, and are on a normal sleep schedule. Sometimes the use of a white-noise generator—such as a fan or humidifier—or earplugs can be helpful for the bed partner. (Applying several layers of duct tape over the sleep-talker's mouth is sometimes a strong temptation . . .) Otherwise, sleeping in another room may be the only solution, which has its obvious drawbacks.

Chapter 16

PARASOMNIA

Shadows of Obstructive Sleep Apnea

I n recent years, it has become more apparent that those with obstructive sleep apnea are far more vulnerable to parasomnias than previously recognized. Considering that OSA is on the rise, it's important to understand what other serious disorders may tag along.

In part, it's the OSA itself that can bring out parasomnias, but the treatment of OSA may also trigger other disorders.

SLEEP TERRORS

A 35-year-old man was in a Pennsylvania hospital for evaluation and treatment of pneumonia, when the house staff and nursing staff noted that he was a loud snorer and seemed to have obstructive apneas during sleep.[1] A brief test and a continuous positive airway pressure (CPAP) trial confirmed OSA, and after the man was released from the inpatient unit, he went to the hospital's sleep lab to have further testing to determine the frequency and severity of his OSA and find an appropriate CPAP level.

There, the staff found that he had more than 73 apneas and hypopneas (breathing that is too shallow or slow) per hour, leading to a diagnosis of severe OSA. Once they determined this, the next order of business was to find out the minimum effective level of pressure needed for the CPAP to be effective. The first level they tried produced only minor improvements, so they increased the pressure. Immediately, there was dramatic improvement in his breathing and sleep quality. He entered deep sleep quickly and had very few apneas and hypopneas. But just as one apnea ended, he suddenly became aroused, clutched his mask, and began thrashing around. He jumped out of bed screaming and pulling off his electrodes and other devices. For ten minutes, his heart rate soared from 71 beats per minute right before the episode to more than 120 beats per minute, while he screamed uncontrollably and knocked a picture off the wall.

When the technician came in, the man seemed disoriented and didn't remember screaming, though he had a vague sense that someone was attacking him and he had to protect himself. He had a history of mild sleep terrors in his childhood and young adulthood, but reportedly, nothing like what he experienced in the sleep lab. Usually, he would just cry out and not leave the bed. His wife noted that he would often thrash around in bed, sit up suddenly, and occasionally sleep-talk, though he never showed any signs of fear or anxiety during these behaviors. These were likely confusional arousals rather than "true" sleep terrors. However, both are classified as disorders of arousal with the same underlying causes.

The technician continued the nasal CPAP titration test and raised the pressure to the satisfactory level without any further problems, and the man was then prescribed a home CPAP unit. Six months later, he was using the unit every night and his wife reported no episodes of sleep terrors, thrashing around, or suddenly sitting up in bed and talking.

The doctors came to an interesting conclusion: The sleep terror occurred because the man had an apnea from unusually deep sleep—"rebound" delta sleep. It's likely that he hadn't had deep sleep in a long time due to his extreme OSA. His sleep was disrupted so often that it's likely he rarely made it far enough to enter NREM stages 3 and 4.

PSEUDO-RBD

Every now and then, OSA can look like RBD. At the sleep clinic at the University of Barcelona, Spain, 16 older male patients who appeared to have RBD didn't—they actually had severe OSA that mimicked RBD.[2] They complained of harmful or potentially harmful dream-enacting behaviors and bad dreams, in addition to snoring and excessive daytime sleepiness.

But on PSG study, it was shown that all of the patients had atonia—that is, they were not able to move—during REM sleep, which is as it's supposed to be. They also didn't have muscle twitching during REM sleep, so no RBD. But they did have severe OSA (and severe obstructive hypopneas), with an average of 67.5 respiratory disturbances per hour.

Two patients had assaulted their spouses during sleep, five fell out of bed, and two suffered lacerations on the face and arms. All reported having dreams similar to what might be expected in RBD, such as being attacked or chased by humans or animals.

When doctors examined the videotaped PSGs, they saw kicking, gesturing, arm-raising, shouting, and talking. These behaviors always coincided with an arousal from REM or NREM sleep triggered by an apneic event. Every time the patients had a dream-enacting behavior, they had just stopped breathing and were aroused by the need to breathe. These behaviors did not arise without that trigger.

Of the 16 patients, 3 refused treatment and 13 were treated with CPAP. The 3 who refused treatment continued to have dream-enacting behaviors and excessive daytime sleepiness. The 13 who were treated with CPAP had their abnormal sleep behaviors, bad dreams, snoring, and excessive daytime sleepiness completely eliminated. Follow-up PSG studies using CPAP confirmed that their sleep-disordered breathing was under control, and also confirmed once again the absence of RBD.

Since OSA is a common sleep disorder that is most frequent and severe during REM sleep, when RBD emerges, this form of "pseudo-RBD" parasomnia may be considerably more prevalent than currently recognized.

SLEEPWALKING

You may remember the horrific story of the woman who cut up her cat on a cutting board, which I mentioned in chapter 3. She was 54 years old and had been experiencing parasomnia events for five years. It started when she was taking a nap in the car while her daughter went grocery shopping, and the woman awoke in a busy intersection three miles away. After that, about five times a month, she'd have experiences of sleepwalking or sleep-driving. Police once stopped her when she was wandering in another town. She was confused, but she awoke in the squad car and was able to tell them where she lived.

This went on for five years before she sought medical attention, and she did so only because her daughter convinced her to, after the woman awoke one morning to find dried blood on her hands, blood on the cutting board, and her cat's mangled body next to the trash. She had been watching a horror movie that night before sleep.

In the PSG study, doctors determined that she had severe OSA with significant oxygen desaturations (not enough oxygen saturating her red blood cells in the bloodstream). They treated her with CPAP, and four months later, she reported that her parasomnia behaviors had completely stopped.

At a hospital in Rhode Island, a 33-year-old obese man with no stated history of sleep terrors or sleepwalking came in for evaluation of his OSA.[3] He had 106 obstructive apnea events per hour, up to 60 seconds long apiece, in the first 2.5 hours of the study. He had severe oxygen desaturations, and no slow-wave (deep) sleep. In fact, his sleep was primarily NREM stages 1 and 2 (light-intermediate sleep), with hardly any REM sleep at all.

Technicians fitted him with a mask and began adjusting the CPAP pressure levels to find the optimal setting. Similar to the patient who had a sleep terror while being treated with CPAP for the first time, this man had a sleepwalking incident once an effective pressure level was found. His apneas ended, and he'd had significant amounts of slow-wave sleep when he abruptly sat up, then walked across the room. He was documented as being in delta NREM sleep at the time. He had a second sleepwalking incident that night that also occurred during delta NREM sleep.

Most likely, this is because during the first night of CPAP therapy, there can be intense slow-wave sleep rebound that instigates sleepwalking. The patient continued CPAP treatment at home, and had no more sleepwalking episodes.

Chronic sleepwalkers who haven't found success with medication, relaxation techniques, or hypnosis may want to be evaluated by a sleep doctor and have a PSG done with a careful eye cast on any sleep-disordered breathing. While not all of those with both sleepwalking and OSA will find that CPAP or surgery for OSA will put an end to sleepwalking episodes, it does work for some.

An OSA Dream

This is an extraordinary dream report with disordered arousal written by an intelligent and articulate 47-year-old woman with OSA and lifelong sleepwalking. She is a filmmaker who, in the sleep lab, had sleepwalking and other abnormal arousals initiated by OSA events.

In my dream, I receive a set of leather boxes as a gift. They are rich cordovan. One box contains a folded brown suit (man's). I see that the suit is too large and I am disappointed. A tall, blond woman wearing a white rayon jumpsuit and diamond jewelry marches past us and jumps from the pier (or dives). I notice, but I am too taken with the boxes and their contents. I think, *Why is this suit brown? I don't wear brown.*

There seems to be a party at some back tables and I realize it is very late, late night and a man is entertaining at his restaurant in Chicago. We have small dishes of pasta and caviar. I walk over to watch a television, or is it an aquarium? There is a picture of a woman's hand, underwater. The nails are French manicured and painted coral pink. They cover almost the entire screen. The colors—aqua, pink, and green (of seaweed)—are brilliant. I notice how well lit and beautifully composed the picture is, and I think, *This is not video format.* Since the lighting is so bright, I see that the woman must be in shallow water on a sunny day, easily seen from the surface (unless the sun is too bright). The hand brushes, or perhaps just appears to brush, a slug or leech from an arm.

I see this open mouth of the woman (the one who earlier jumped) buried in the sludge of the bottom of the lake or sea. Her eyes are open. She is motionless. I am mesmerized by seaweed and strands of hair swaying with the current. I think, *This scene is too long.* Suddenly I realize that this is not TV nor an aquarium. This is reality and I am underwater. I realize that I will never get back to the surface in time. I awaken gasping for air.

SLEEP BRUXISM

OSA also puts people at higher risk for sleep bruxism (teeth-grinding). In a small study from the University of British Columbia, 21 randomly selected patients with OSA were given questionnaires and studied overnight with PSG.[4] Six out of 11 patients (54 percent) with mild OSA and 4 out of 10 patients (40 percent) with moderate OSA were diagnosed with sleep bruxism.

However, the researchers note that the timing of the teeth-grinding events rarely matched the apnea episodes. They concluded that the sleep bruxism wasn't directly related to the apnea events but was related to the generally disturbed sleep of OSA patients.

PERIODIC LIMB MOVEMENTS AND RHYTHMIC MOVEMENT DISORDER

People with OSA have a higher incidence of periodic limb movements (PLM) during sleep, although the prevalence is not well established, and, more important, the clinical relevance is quite uncertain. PLMs can appear before, during, or after OSA events. Also, PLMs can emerge with control of the OSA. Once OSA is controlled with CPAP, if PLMs persist or emerge anew and disrupt the sleep of the patient or the bed partner, then treatment with a dopamine-enhancing medication may be useful.

Rhythmic movement disorder (RMD), featuring body rocking and head banging, can also interact with OSA and CPAP treatment in similar ways as PLMs. The so-called central pattern generators in the brain stem and spinal cord that generate rhythmic and non-rhythmic simple movements or complex behaviors can become activated and deactivated with sleep-related breathing disturbances and their reversal with treatment. In other words, the breathing and motor-behavioral systems in the brain and body interact in complex ways during sleep, disturbed sleep, and the control of disturbed sleep.

The important point is to be aware of any new or persistent changes in sleep behaviors associated with sleep-disordered breathing and its therapy, and to then discuss these changes with your sleep doctor.

SEIZURES AND
ANOXIC ATTACKS

OSA can shock the brain and induce epileptic seizures with complex or violent behaviors, or nocturnal cerebral anoxic attacks with vigorous behaviors. They can look alike, so it's difficult to differentiate between seizures and cerebral anoxic attacks without an EEG monitor while it's happening.

One cerebral-anoxic-attack case report does have PSG documentation, however. A 52-year-old factory worker visited doctors because of daytime sleepiness and heavy snoring. The previous year, he'd gone into a coma that lasted several hours during sleep because of too much carbon dioxide in his blood. Clinicians discovered that this man had amazing apnea episodes—many lasting more than three minutes!—during REM sleep. During most of these episodes, his EEG activity would slow down and then flatten, then he'd have a generalized spasm that lasted 5 to 10 seconds. He remembered none of this in the morning.

Because of the life-threatening danger of his condition, the man underwent a tracheostomy, which is a surgical opening created in the lower windpipe to allow for breathing below the upper airway obstruction that occurs in sleep. He also underwent uvulopalatopharyngoplasty surgery, during which excess tissue is removed from the back of the throat and the soft palate to widen the airway passage. Three months later, he came back for another PSG study, which showed regular breathing while he was lying on his side, and continuing apneas when he was lying on his back—but they were shorter (under 40 seconds), and no attacks occurred.

In the case of seizures, OSA and the seizure disorder can feed off each other.[5, 6] OSA can aggravate the seizures because of sleep fragmentation, low oxygen levels, and decreased cerebral blood flow. Sleep deprivation is known to be one of the leading triggers of seizures in those with a seizure disorder. Understandably, treatment of OSA often reduces seizure frequency and severity in those with seizure disorders. In rare cases, epileptic seizures have been misdiagnosed as OSA, in one case leading to a lack of treatment for epilepsy for 52 years!

OTHER PARASOMNIAS
FOUND WITH OSA

All varieties of arousal disorders are possible from OSA-induced arousals. Also, instinctual behaviors—sex, eating, locomotion, aggression, violence— often appear during OSA-related parasomnias.

Sleepsex is known to occur with OSA—sometimes it's even reported that the man was snoring (and definitely asleep) while having sex. In these cases, when CPAP was used to control OSA, it also controlled the sleep-sex in all reported cases. Sleep-related eating disorder (SRED) can also spring from repeated episodes of OSA. Again, control of OSA with CPAP will usually also stop the sleep-eating. However, some patients continue to eat despite control of their OSA, in which case separate treatment with bed-time topiramate (a first-line treatment of SRED) is necessary.

In a curious example of the link between OSA and sleep-eating, there was the reported case of a man who gained enough weight from SRED to push him into developing OSA, and when he eventually lost weight, then the OSA also subsided. How did this man lose all that excess weight? He was serving time in jail. Of course, when he was released, he gained the weight back and back came the OSA. He refused CPAP treatment.

Violence during confusional arousals is also possible with OSA—such as with the man mentioned earlier who accidentally grabbed a gun and shot himself in the head, thinking he was attaching his CPAP mask. Fortunately, it was a tangential gunshot wound to the scalp and he lived. People may become violent to their bed partners, or they may sleepwalk outside the home and become violent to people they encounter. All of this under the influence of . . . OSA.

PREVALENCE STUDIES

There have been several studies to examine the relationship between OSA and parasomnias. In each case where surveys are used, the question-naires are slightly (or very) different and different criteria are used. There-fore, the numbers can vary—but we can see a strong relationship between OSA and parasomnias by looking at the overall picture.

Children's Studies

The largest general-population-based (as opposed to clinic-based) study of parasomnias and sleep apnea in children involving PSG monitoring was published in 2004. Researchers recruited children 6 to 11 years old, and ended up studying 480 PSGs and corresponding parent questionnaires.[7]

For the purposes of this study, the researchers classified something as a parasomnia only if it happened at a minimum frequency. For example, the parent had to say that the child had sleepwalking episodes at least three times per month for the researchers to classify the child as a sleepwalker; the child had to reportedly talk in his or her sleep at least five times a month to be called a sleep-talker; and the child had to have at least five fearful awakenings a month to be classified with sleep terrors. Enuresis (bed-wetting) had to occur at least five times a month to qualify. For other symptoms—such as insomnia, excessive daytime sleepiness, and snoring, they counted only if they were reported as occurring "frequently" or more.

Children who had at least one apnea or hypopnea per hour of sleep were found to be more likely than the other children to sleepwalk (7 percent versus 2.5 percent), sleep-talk (18.3 percent versus 9 percent), or wet the bed (11.3 percent versus 6.3 percent). They also were more likely to have learning difficulties. Even just *one* apnea per hour makes this much of a difference! Keep in mind that some people have more than 50 apneas per hour.

Children who experienced two apnea episodes per hour of sleep were three times more likely to have sleep terrors than the others.

The researchers noted that children often had more than one arousal parasomnia (sleepwalking plus sleep-talking, sleep-talking plus sleep terrors). Those with parasomnias were also more likely to have insomnia, excessive daytime sleepiness, and learning problems.

In a study at Stanford of prepubescent children who had been evaluated for treatment of sleep terrors with or without sleepwalking, the researchers found that 49 of 84 children with sleep terrors had a diagnosis of sleep-disordered breathing confirmed by PSG.[8] Of them, 43 were treated surgically for the sleep-disordered breathing, and in each of these cases, parents reported that it eliminated the sleepwalking and sleep terrors—which leads to the conclusion that the sleep-disordered breathing is what triggered the parasomnia (in genetically susceptible children).

Finally, it has been known for ten years that children with obstructive sleep apnea are prone to parasomnias.[9]

Adult Studies

In one study, 4,972 adults in the United Kingdom were interviewed by telephone.[10] The researchers were trying to find out what factors were associated with the arousal parasomnias.

The two factors that were most often identified by people with sleep terrors and sleepwalking were "a subjective sense of choking or blocked breathing at night"—which may be a sign of OSA—and obstructive sleep apnea syndrome.

FINAL WORDS ON OSA

One of the many reasons that OSA is such a troublesome disorder is that so many parasomnias can follow along with it. It's essential that anyone with symptoms of OSA have a clinical sleep evaluation and an overnight PSG study. In many cases, treating the OSA also controls or improves the "tag-along" disorders.

Chapter 17

PARASOMNIA POTPOURRI

Nocturnal Groaning, Exploding Head Syndrome, Sleep-laughing . . . and Pets with Parasomnias

Some parasomnias, while interesting and strange, don't usually cause problems or generate complaints and therefore haven't been the subjects of a great deal of medical research. Nevertheless, they're among the disorders that can arouse much curiosity. We'll discuss some of them here.

SLEEP-LAUGHING

Laughing during sleep is usually not considered a problem, except for the potential it has to disrupt a bed partner's sleep. Some people report that it relates to their dreams—they'll remember the laughter as relating to a silly or funny dream—while others say they have no idea why they were laughing and don't connect it to any dreams. Sometimes sleep laughers may be completely unaware of their laughing, which usually occurs in the second half of the sleep cycle during REM sleep or in the transition from REM sleep to partial arousal or wakefulness.

A study in 2005 by the Sleep and Alertness Clinic in Toronto studied a small group of sleep-laughers.[1] A few of them also experienced sleep-talking. The frequency of sleep-laughing episodes ranged from once or twice a week to once every few months.

Interestingly, the researchers wrote that patients tended to remember sleep-laughing much more than sleep-talking. They also normally remember the dream they were having, often describing it as bizarre. And most of the patients said that the same story wouldn't make them laugh if they were awake.

A rare cause of sleep-laughing is sleep-related *gelastic seizures,* which are laughing seizures that occur predominantly during wakefulness. Epilepsy can be "the great imitator" of all the parasomnias, including sleep-laughing.

Callie's Story

I was awakened last night by one of my sleep-laughing fits. Weird thing. I'll wake up laughing *out loud,* feel amazingly happy and goofy from the laughing, and then start to worry about my sanity. This little nighttime tendency has been going on since I was a kid. My mom used to hear me laughing from my bedroom when we lived in my childhood home.

I guess that means some episodes *don't* wake me. My partner, Kev, has been known to stroke my back to wake me from a few of my sleep-laughing spells. He didn't hear last night's. It must have been a doozy to have awakened me, though.

I never seem to be able to associate the ones that wake me with precipitating dreams. No mental scenes or scenarios seem to precede the laughing fits. My body/mind just decides to let loose with a laughing spiel.

Don't know if any of it means anything. I laugh plenty when awake. It's not as if I'm below quota. Maybe it's a stress thing. Maybe it's my body's way of releasing endorphins. I have been a bit stressed the past week or so. Hmmmm. I guess I shouldn't complain. It could be worse!

"You have been dreaming at a great rate—talking in your sleep, laughing, and clapping your hands as if you were cheering some one. Tell me what

was so splendid," said Mamma, smoothing the tumbled hair and lifting up
the sleepy head.

—from "A Christmas Dream, and How It Came True,"
The Louisa Alcott Reader, Louisa M. Alcott

Multiple Sleep Behaviors with Laughing

Tracey Hoyng says that she knows to expect weird sleep behaviors
when she's been stressed. She doesn't remember these behaviors, but her
husband tells her about them. "My husband usually awakes to find me
laughing hysterically, different from my 'normal' laugh. He'll find me
waving at someone who's not there, and often carrying out a one-sided
conversation with God knows who," she says.

He wakes up startled, sometimes thinking Tracey is awake and talking to him, but he figures it out quickly, knowing her history of sleep
behaviors.

"He says that most times I tend to laugh or giggle, other times I say 'Hi'
and wave, and sometimes I'll say things like 'Really?' or 'Nah, that's okay,'
as if I am talking to someone. I usually calm down after a few minutes, and
I tend to wake up exhausted the next morning, on these occasions especially," she says.

EXPLODING HEAD SYNDROME

An episode of exploding head syndrome would feel something like this:
You're lying in bed, about to go winging off to slumberland, when you
hear a great noise, like a pair of giant cymbals clapped behind your head.
Except not exactly behind your head. More like *in* your head. It jolts you
awake, and you look to and fro for a second, but the room is silent. No
cymbal-clappers to be seen.

You start to drift back off when it happens again. A car backfires in your
ears. Once again, you sit up, every hair on your head and body erect, heart
thumping in your chest, breath coming in gasps. It feels like your brain is
a nuclear blast site.

That sliver of hazy consciousness between the waking state and sleep is full of strange occurrences. Many people experience hypnic jerks, and some feel as though they're falling, a sensation often connected to vibratory noise.[2, 3] Most EHS episodes occur in this murky transitional period, "the twilight state of sleep."

EHS, first described in sleep literature in 1988, is a condition in which subjects hear bursts of sound in their head or see flashes of light, usually as they're dropping off to sleep. People describe them as "electrical head shocks," "brain blasts," and plenty of other colorful things. They compare the sound to gunshots, thunder, loud buzzing, something hitting the window, firecrackers, exploding bombs, and pounding drums. Some people perceive the noises as internal (happening inside their bodies), while others perceive them as external (happening in the house or the street outside). The sound is usually described as very loud, but the volume and intensity vary.

The experience often comes in waves: It may happen every night for weeks, then disappear for months at a time (or forever). There doesn't appear to be a strong correlation with other sleep disorders, psychological disorders, or medical conditions.

Nobody is certain why these attacks occur, but they are considered benign (not dangerous). They do not cause subjects pain, but they can cause intense anxiety and poor-quality sleep. EHS is *not* a headache disorder, it is a sensory parasomnia.

The tricyclic antidepressant clomipramine (Anafranil) can be used to reduce the severity of EHS, along with other medications, such as the antihypertensive nifedipine (Procardia, Adalat, Nifedical). Most people who suffer from EHS need only be assured their condition isn't a threat to their health. This is important information to share, since sufferers of EHS may be concerned that they have a brain tumor, stroke, or some other major brain problem.

A published report from 1991 followed nine individuals who complained of EHS. None had histories of neurological or epileptic disorders. During polysomnographic recording, five of them had EHS episodes while they were being tested, during times when they were relaxed and awake, trying to fall asleep. The severity of symptoms eased as the subjects learned that the condition was harmless. Clomipramine was given to three subjects, and they reported that the drug alleviated their symptoms.

SLEEP-HUMMING: EZRA POUND

Poet Ezra Pound was hospitalized for 12 years for being psychotic (schizophrenic), which he was not, but better to be there than to be in prison or executed for treason during World War II. According to the biography *A Serious Character: The Life of Ezra Pound,* the night nurses at St. Elizabeth's Hospital heard him humming all night long—which certainly sounds like Mr. Pound was a "sleep-hummer." Any sort of vocalization you can make while awake, you can also make while sleeping.

NOCTURNAL GROANING

Most of us make little noises in bed as we sleep or during mini-arousals: grunts, snorts, mumbles, murmurs, grumbles, yawns. But some people go far beyond those sounds. We're talking beyond snoring, beyond sleep-talking. We're talking about nocturnal groaning, also known by its Latin name, *catathrenia.*

Catathrenia is a disorder characterized by groaning during sleep, especially during REM sleep in the second half of the sleep cycle. Whereas snoring happens during inhalation, catathrenia happens during exhalation, and can be loud and prolonged. This is a chronic (long-lasting) disorder, identified as a parasomnia in 2001.[4] In that case series, catathrenia was associated with episodes of bradypnea, an unusual slowing in a person's breathing rate that mainly occurred during expiration in REM sleep.

Catathrenia can occur in clusters throughout the night, and usually occurs every night. It's often disturbing and creepy to bed partners and others in the vicinity. In a study of four patients with the condition, the groaning sounds lasted from 2 to 20 seconds apiece, and the episodes lasted from 2 to 60 minutes, recurring several times throughout the night. This condition began in childhood or adolescence in all four patients.

Although the noises may sound miserable, or even "mournful," they are not apparently emotion-related. The person does not show discomfort or a tense facial expression, usually lies still, and doesn't feel excessive daytime sleepiness or have a depressed mood as a result. There is no awareness of the groaning by the groaner, which is similar to what happens with snoring.

Unfortunately, despite multiple interventions, there is no known treatment yet for this condition, although in at least one published case report, a woman with both nocturnal groaning and obstructive sleep apnea was able to control both conditions with CPAP therapy.[5]

HICCUPS AND SLEEP

Hiccups that persist into sleep enter the realm of parasomnias. Doctors in Paris published a report about the sleep effects on hiccups that had been present for anywhere from seven days to seven years in eight patients who complained of insomnia and who were studied overnight in the sleep lab. Hiccups were most common during wakefulness and then became progressively less frequent as sleep deepened all the way to delta NREM sleep and REM sleep. Of the 21 bouts of hiccups they saw stop, half occurred during episodes of sleep apnea or hypopnea (reduced airflow). Hiccups themselves did not cause any arousals from sleep, but they hindered the patients as they tried to fall asleep. These patients slept poorly, with long waking periods. The authors of this study concluded that the brain mechanisms responsible for hiccups (which involve spasms of the diaphragm and chest muscles), presumably located in the brain stem, are strongly influenced by sleep and breathing disorders of sleep.

PROZAC EYES

Prozac is almost certainly the most iconic psychiatric medication. It's a buzzword, a household name, and it's in the title of books. There is no doubt that Prozac (fluoxetine, its generic name) has been highly effective in controlling depression in many treated patients. The same holds true for all the other SSRIs now available. Selective serotonin reuptake inhibitors (SSRIs) are the chemical class of antidepressant for which Prozac was its pioneer beginning in the mid-1980s. Prozac is usually well tolerated, although known side effects include insomnia, nightmares, sexual dysfunction, and weight loss or gain.

There is also another recognized type of side effect from Prozac therapy—the sleep-motor and behavioral disorders. Prozac and all the other SSRIs, along with venlafaxine (Effexor), mirtazapine (Remeron), tricyclic antidepressants, and a few other drugs, can induce or aggravate REM sleep behavior disorder, restless legs syndrome, and periodic limb movement disorder. Prozac can powerfully affect the movement of the eyes during sleep—a phenomenon known colloquially as "Prozac eyes." "Prozac eyes" refers to prominent, sustained, and high-voltage eye movements in the transition from being awake to being asleep, and also in light NREM sleep.[6] The eyes move with slow or medium speed.

People who go through this don't complain that their eyes hurt or that it wakes them up or that they have any discomfort upon awakening in the morning. In fact, to date there is no known medical problem from having Prozac eyes. It may also be a permanent side effect, since one patient was studied 17 months after he had stopped Prozac, and he still had Prozac eyes—and still had RBD that originally had been triggered by Prozac. Many other patients with Prozac eyes have reported stopping Prozac or another SSRI months or years before their sleep-lab studies.

Prozac eyes indicate that serotonin has a strong influence in the brain stem, where eye-movement activity is generated and modulated. (Another cause of Prozac eyes is brain-stem disease, especially neurodegenerative disorders.) Also, Prozac is known to trigger or aggravate RBD, which indicates its potent and disturbing effect on brain-stem–motor-behavioral activity in REM sleep. This is a major and frightening consequence.

In the sleep lab, it's easy to tell when patients are on these antidepressants—or have been exposed to them in the past—because of their eye movements.

NIGHTMARES

Now officially termed "nightmare disorder" in ICSD-2, "bad dreams" involve emotionally disturbing dreams that often awaken the sufferer, who can immediately recall the content of the dream and its associated negative emotions. It is now known that not only fear, but also a variety of other negative emotions, such as disgust, can be present during a nightmare. A genetic basis for nightmares has been demonstrated, with personality factors, and drug and alcohol use playing mediating roles. Of course, traumatic events (such as assault, war experiences, etc.) are well known to trigger recurrent nightmares, which can be very persistent and highly distressing. However, in some cases there can be an adaptive component to nightmares, as when a person is "working through" a problem while dreaming.

Fortunately, both psychotherapeutic[7, 8] and pharmacologic[9, 10] therapies can be helpful in reducing or fully suppressing nightmare intensity and frequency, although more research needs to be conducted. There are various cognitive-behavioral therapies that have been developed and that can be effective in six or fewer sessions, which use systematic desensitization and relaxation techniques in order to induce relaxation and mental restructuring in response to anxiety-provoking nightmare content. For example, "imagery rehearsal" involves the technique of changing a recalled nightmare and rehearsing new scenarios. Hypnosis offers some promise. In terms of pharmacologic therapies, in my experience I have found greatest success with cyproheptadine, prazocin, and clonidine taken at bedtime, but not uncommonly I have to "go down the list" of other options. Of course, combined psychotherapeutic and pharmacologic interventions may be the best option for some patients.

Finally, we should keep in mind that various medications, such as SSRIs, can induce vivid dreaming and nightmares, along with a large number of medical, neurological and psychiatric disorders.

PETS WITH SLEEP DISORDERS

Humans aren't the only ones who experience sleep disorders; others in the animal kingdom are also vulnerable to some of the same sorts of problems. Domestic animals have been found to have what appear to be sleep terrors, insomnia, bed-wetting, limb movements during sleep, violent behavior during sleep, nocturnal seizures, and other behaviors similar to their human counterparts.

Narcolepsy in Dogs and Horses

Narcolepsy has been documented in dogs and horses. In particular, Doberman pinschers, Labrador retrievers, poodles, dachshunds, and beagles are known to be vulnerable to narcolepsy. The affected dogs tend to have cataplectic attacks whenever they are excited. Dogs tend to get very, very excited about food . . . so feeding time can lead to immediate collapse.

The consequences of this don't necessarily sound so bad until you hear a story like Skeeter's. Skeeter is a toy poodle with narcolepsy, and the *Idaho*

State Journal reported in 2005 that his owners could no longer feed him dog food because he would consistently collapse on the sight of his food. So they began feeding him grilled vegetables and lunch meat by hand, holding up his hind legs and massaging his neck to keep him alert.

Dogs with narcolepsy tend to be sleepy during the day. The disorder usually presents itself when the dog is a puppy, during the first few months of life. It's treated in the same way as human narcolepsy: with stimulants and antidepressants, typically.

Although the disorder looks similar in both dogs and humans, it's caused differently. Stanford researchers discovered that a single recessive gene causes narcolepsy in Doberman pinschers and Labrador retrievers. This means that both parents must have the disorder or be carriers of the disorder for a dog to become narcoleptic. With other breeds (and with humans), it's not always inherited in this clear manner. Narcolepsy also runs in families in Shetland ponies.

RUSTY, THE NARCOLEPTIC DOG

A short video of a dachshund made its rounds on Internet humor websites recently. It shows the dog running through a field and promptly collapsing from cataplexy while a narrator describes narcolepsy. You can view the video clip at www.devilducky.com/media/8232, and the "music video" version at www.devilducky.com/media/8653.

The following story was told to me by a woman named Carolyn who adopted Sam the Beagle, who suffered from parasomnias.

Sam the Beagle's Story

Driving to the meeting of the local beagle rescue on the morning of February 13, 2005, I was looking forward to learning more about rescue, and how I could help more than the one dog I had fostered the prior month. At about the same time that morning, the family that had adopted a sweet, lovable two-year-old male beagle just two months earlier was putting him into a car. The call I received on the way to the meeting was desperate and unsettling: "Can you please help? Sammy is being returned for biting the father and the little girl in the family. He needs somewhere to stay until we can get him to the veterinarian, and into long-term foster care for assessment. The family will drop him off at the meeting." How could I have said "No, that scares me" when all I wanted to do was help? I reluctantly agreed to bring Sam home.

I had remembered Sammy from the rescue's website in December. He was one of the outdoor hunting dogs from Tennessee who was being cared for by the owner's neighbor. The neighbor was moving, and the owner didn't want the dogs back, so Sam and two other beagles were surrendered to rescue. Sam received routine veterinary care, was treated for heartworms, temperament-tested, and found to be a wonderful family dog who loved children. He was adopted into a New Jersey family, and had been with them only a short time. I was now being asked to bring this "aggressive" dog into my home, so I asked for more information about the biting incidents.

We discovered that, unbeknownst to the rescue at the time, the problem first surfaced two weeks after adoption. Sam was in a deep sleep, cuddled up next to the adoptive father on the couch. Suddenly, Sam started growling, and then quickly whipped his head around and bit the (sleeping) father's arm! They shouted his name, and they report that he seemed to wake up and then just sat there looking at them as if he had no idea what had just happened. The bite did not break the skin, so the family decided it was a one-time incident, and they would let it go.

They said that they would hear Sam snarling and growling in his crate at night. They ignored this as well because no one was ever near Sam at those times, and they never felt threatened by him. He was a sweet little beagle.

Unfortunately, two months later, Sam fell asleep next to the five-year-old little girl while she was watching television. While sleeping, Sam suddenly growled, jumped up with his paws on her shoulders, had his eyes partially open, and bit her on the face. She screamed and cried. Sam stopped, cowered, and went into his crate as instructed by the family members who rushed to the little girl's aid. She was taken to the hospital for bruising, scratches, and two puncture wounds to her face requiring six stitches. A bite report was filed against Sam with the New Jersey Department of Health.

With this history, I realized that the only reason Sam was coming to my home instead of being immediately euthanized was because of the family. They asked that he not be euthanized because "he is not vicious—he doesn't even seem to know what he is doing."

I had Sam for five days while we set up a veterinary screening and a long-term foster home. I saw absolutely no type of aggression whatsoever, although I kept my distance from him, and kept him somewhat segregated from my own two beagles. When the veterinary results came back, the exam and blood work were unremarkable, and medical causes (e.g., a thyroid imbalance) for the reported aggression were ruled out. Sam was then sent to his long-term foster home for assessment. In five days in that home, he bit the foster mother three times. The description of the incidents was the same as those from the adoptive family: Sam fell asleep on the couch next to her, and appeared to be sound asleep when he suddenly growled, quickly snapped, and bit her on the arm. The rescue decided that this was too unpredictable. Sam was deemed "unadoptable" and was to be scheduled for euthanasia.

One of the directors of the rescue and a few volunteers were uncomfortable with the situation—as was I. I felt that if we at least *knew* what was wrong, then maybe we could help Sammy. Having a background in applied behavior analysis, I wrote an actual proposal to the rescue asking for Sam to return to my home for a three-week time period, during which I would ensure everyone's safety, and systematically change different variables in Sam's environment to try to determine the cause of what we called Sam's "attacks." They accepted.

I heard the first of Sam's episodes after about a week in my home . . . in the middle of the night (once at 3 a.m. and again at 4:15 a.m.) while he was sleeping soundly in his crate next to my bed. No one was near Sam, and there were no noises to startle him. These episodes began to average a few times per week and ranged from snarling and growling to this escalation into a bite of whatever was near Sam when he was sleeping (his crate, a pillow, etc.).

I learned that if I called Sam's name when he was snarling, he would awaken peacefully with no escalation into a bite. This ruled out a seizure disorder and confirmed to me that he was truly sleeping during these episodes, and I was simply waking him. I also noticed that these episodes occurred across changes in many variables—diet, sleeping environment, exposure to children, exercise, and so on.

I started researching possible diagnoses online. I consulted with experienced friends in animal welfare, aggression, and epilepsy websites; three different veterinarians; and two veterinary behaviorists. Dr. Amy Marder of the Animal Rescue League of Boston helped me rule out a true behavioral cause for these episodes, and indicated that while she believed there was some type of sleep disorder, she was not very knowledgeable about it, and recommended some local neurologists. I contacted local neurologists, who never responded to my pleas for a discounted consultation (on behalf of the nonprofit rescue organization), and finally stumbled onto work by Dr. Carlos Schenck, Dr. Mark Mahowald, and Dr. Adrian Morrison on REM Sleep Behavior Disorder (RBD).

These researchers confirmed that what I described in Sam sounded like a "classic case" of RBD. Drs. Mahowald and Morrison indicated that although there was a pharmacological treatment (clonazepam), Sam could never be cured, and would probably always need to sleep by himself in a crate. I was hesitant to put Sam on lifetime medication, and decided I would manage the episodes environmentally and pursue medication in the future if the incidents increased in intensity, duration, or frequency.

With this information, Sam was still deemed "unadoptable" through the rescue organization. He ended up staying with me for four months before I finally asked if I could adopt him. I just could not imagine this sweet soul being euthanized for a completely manageable sleep disorder. Sam has thrived in my home, and has won the hearts of my friends, family, and his beagle brother and sister. He still sleeps by himself, but now he's surrounded by a loving environment. Learning about RBD quite literally saved Sam's life. Maybe his story will help save another.

DIAGNOSING SAM

When Carolyn contacted us, she had been fostering Sam for a month and a half, and was searching for a diagnosis. I was struck by the similarities between Carolyn's descriptions of Sam and wives' descriptions of husbands with RBD. In both, the violent behavior during sleep was completely out of character with the wakeful personalities.

Carolyn came to the annual meeting of the New Jersey Sleep Society, where I was giving a presentation about RBD. She spoke about Sam, and showed a video that displayed his sweet wakeful personality then two of his violent sleep episodes. Both of the episodes are extremely short. The dog growls for a few seconds, then half-opens his glassy eyes. Immediately, he turns and snaps his jaws into the air—then whimpers, as if he's just frightened himself.

Sam was very lucky to find an adoptive mom who was as persistent and caring as Carolyn; he was scheduled for euthanasia when she asked to keep him. She has been able to manage Sam's disorder environmentally for more than a year, by being perceptive and making sure no one is nearby when Sam falls asleep. If it happens accidentally, Carolyn knows approximately how long it takes for Sam to transition into REM sleep, and knows that she can wake him by calling his name. He's had no further biting incidents since living with her.

Inspired by Sam and his close escape from death, Carolyn cofounded Cascade Beagle Rescue–East (www.cascaderescue.org). She writes, "In rescue, every beagle, every adopter, and every volunteer has his/her own story. We need to slow down and take the time to listen. There is a solution to every problem, and almost always a way to help if people work together."

Finally, in a nice display of reciprocity, while experimental animal studies have helped us understand RBD in humans, the treatment of RBD in humans in turn has been used to control naturally-occurring RBD in cats and dogs whose owners bring them to veterinary clinics.

IN CONCLUSION

The study of sleep teaches us that for all our vast and rapidly increasing medical knowledge, we're still babes in the woods. It was barely 50 years ago that we even learned that such a thing as REM sleep existed, and now we're continuing to uncover "new" conditions, parasomnias that have likely existed across species since well before the arrival of man.

Maybe someday we'll decode the mechanisms behind those nonsensical ravings during sleep, or be able to explain why some people groan, laugh, cry, or feel like their heads are exploding just before sleep. In the meantime, we will continue to gather more pieces in the sleep-brain-mind-body puzzle and keep trying to put them together in meaningful ways as we explore the mysterious realm of sleep.

Chapter 18

BED PARTNER'S INSOMNIA

*Sleeping with a Snorer,
Leg-jerker, Loud Sleep-talker,
or Parasomniac*

You might be a person who practices great sleep hygiene and has no predisposition to sleep disorders, but you just happen to fall in love with a person whose sleep mannerisms seem designed to ensure that you'll never get a good night's sleep again. It's so unfair. Or even less fair: You may not even be in love with that person. It may be your college roommate, sibling, or anyone else sleeping in the vicinity.

It's officially considered an environmental sleep disorder if your sleep is disrupted due to a bed partner's behaviors or to other problems associated with your sleep environment—such as annoying sleep sounds coming from your roommate, loud street noise, temperature that's too hot or cold, flashing lights, and so on. The lights, noise, and forced awakenings associated with hospital stays are known to cause sleep disruption.

It's also considered an environmental sleep disorder if you're forced to be ever alert to care for an infant or a sick or elderly person who may need attention throughout the night. That one gets a special name: *caretaker insomnia.*

PRIMARY AND SECONDARY
DISORDERS

In some cases, the "disruptive" person has a diagnosable sleep disorder, and in others, the behavior is not classified as a sleep disorder but is nonetheless disruptive.

For example: snoring, by itself, isn't a sleep disorder. It may be a sign of a sleep-related breathing disorder such as OSA, but it doesn't have to be. Regardless, snoring can certainly disturb a bed partner's sleep, causing insomnia (a diagnosable sleep disorder). The insomnia may be sleep-onset insomnia (trouble falling asleep) or sleep-maintenance insomnia (trouble staying asleep).

Of the movement disorders, RLS and PLMD are the two that cause a bed partner's insomnia most often. The person with the movement disorder would be said to have a *primary* sleep disorder, and the affected bed partner's insomnia would be considered a *secondary* sleep disorder. That is, it should resolve if the primary sleep disorder is solved.

But many of the other types of behaviors that we've discussed in this book—sleepwalking, sleep-laughing, nocturnal scratching, teeth-grinding, etc.—can cause enormous problems for spouses or bed partners.

CARETAKER INSOMNIA

It's such a well-known phenomenon that it becomes almost a joke: doctors always advise new parents-to-be to get lots of sleep in the weeks leading up to the mother's due date. The thinking is, of course, that they're probably not going to get uninterrupted sleep again for quite a while once the baby is born.

This same phenomenon occurs among those who care for people who have health problems, dangerous sleep disorders, or who experience multiple awakenings through the night. In these circumstancs, their own sleep becomes unpredictable and insufficient as a result.

Some people are able to take naps during the day and have mild or no daytime consequences. But many others develop the signs of excessive daytime sleepiness, such as loss of concentration and attention, irritability, and depression.

A doctor can prescribe a low dose of a sedative (such as 0.125 or 0.25 mg of alprazolam, or 0.5 mg of lorazepam) at bedtime to help the person return to sleep quickly after multiple awakenings throughout the night.

Naps of 15 to 45 minutes and the use of caffeine to stay alert in the morning can both be helpful. However, prolonged naps or naps taken late in the day could aggravate insomnia, as could excessive amounts or ill-timed use of caffeine. Regular exercise during the day is also beneficial, though it's not recommended at night.

NOT JUST INSOMNIA

Sometimes a partner's sleep disorder can cause psychological distress that manifests beyond insomnia. In one notable case, a 27-year-old woman developed PTSD after her husband had a severe sleep-terror episode in which he jumped out a closed second-story window.[1] He suffered many lacerations but clung to the roof and didn't fall. She was startled awake and screamed to him to try to alert him and coax him back inside.

After that incident, she developed extreme anxiety and would no longer sleep in the bedroom—instead, she slept on the floor in a different bedroom for a year. She reported having frequent nightmares and hyper-vigilance, and a sense of detachment from her husband. After cognitive behavioral therapy, her symptoms improved, she was able to sleep with her husband again, and both partners reported that their relationship was improving.

THE SLEEPLESS HUSBAND

A 35-year-old woman came to see me with her husband. She had been having very frequent sleep terrors, up to eight per night, that started shortly after they got married and kept increasing in frequency. Although she didn't remember her sleep-terror episodes, her husband always awoke and tried to calm her down. Sometimes she injured him—she dug her fingers into his eyes until he developed blood spots, and pulled his hair. She also once went into the children's bedroom and carried her daughter out into the hall. Because of the frequent sleep interruptions, and the constant worry that she might harm the kids or herself, the husband had developed sleep-maintenance insomnia.

Even on the rare nights when she didn't have sleep terrors, "I don't sleep so good anymore," he told me. He was just expecting to be interrupted and had become a much lighter sleeper. Then he felt wiped out all day. He described himself as "understanding, but worn out."

Sleep habits die hard. It can take some time after the primary sleep disorder is controlled before the bed partner can get back to a normal night's sleep.

SOLUTIONS

Unfortunately, when the bed partner's insomnia gets bad, couples sometimes resort to sleeping in separate beds—or even separate rooms, which can interfere with intimacy. But there are several measures to try first.

If the problem is noise (snoring, bruxism, groaning, etc.), try wearing earplugs or using a white-noise machine or fan to drown out the sound. Soft instrumental music might work, too. If your partner's noises prevent your ability to fall asleep, you should try to go to sleep before your partner does. Get a head start while it's quiet in the bedroom!

Find out if there are certain positions where your partner tends to snore more. Often, lying on the back is the worst position. Encouraging your partner to sleep on his or her side or front may help. You can try to "hold" this position in place with pillows or your own body. Also, special nightshirts are available, or can be made, in which a pocket sewn to the back allows the placement of a tennis ball, so that the snorer finds it too uncomfortable to sleep on his back and will then shift to his side or stomach as the preferred sleeping position, which will then reduce the snoring. Your partner may also try using an extra pillow—or removing the extra pillow.

There are many devices intended to reduce snoring—nasal strips, devices inserted into the nose, appliances fitted to the teeth to bring the jaw forward, palatal implants, and so on. Anyone who is a loud snorer should be tested for sleep apnea; after the exam, you can find out more about snoring treatments to try.

If the problem is movement (restless legs or periodic limb movements), you may try a bigger bed or try placing a body pillow between the two of you.

And, of course, seek professional consultation if you suspect that your bed partner has a sleep disorder. The fact that it's disruptive to *your* sleep may be enough impetus for your partner to seek help, even if he or she isn't bothered by the behaviors.

Glossary

These acronyms have become part of our clinical notations and are used in sleep-journal publications.

APAP: auto-adjusting positive air pressure
BiPAP: bilevel positive airway pressure
CA: confusional arousals
CPAP: continuous positive airway pressure
CRD: circadian rhythm disorder
DD: dissociative disorder
DID: dissociative identity disorder
EEG: electroencephalogram
EHS: exploding head syndrome
EKG: electrocardiogram
EMG: electromyogram
HH: hypnagogic (HG) hallucinations/hypnopompic (HP) hallucinations
ICSD: international classification of sleep disorders
MSLT: multiple sleep latency test
NCPAP: nasal continuous positive airway pressure
NES: nocturnal eating syndrome
NOC DD: nocturnal dissociative disorders
NOC SZ: nocturnal seizures
NREM: non-rapid eye movement

OSAS: obstructive sleep apnea syndrome

PLMD: periodic limb movement disorder

PLMS: periodic limb movements during sleep

POD: parasomnia overlap disorder: RBD/SW/ST

PSG: polysomnography

RBD: REM sleep behavior disorder

RDI: respiratory disturbance index

REM: rapid eye movement

RLS: restless legs syndrome

RMD: rhythmic movement disorder

SBS: sexual behaviors during sleep

SOREMP: sleep-onset REM period

SP: sleep paralysis

SRE: sleep-related erections

SRED: sleep-related eating disorder

ST: sleep terrors

STATUS D: status dissociatus

SW: sleepwalking

SWS: slow-wave sleep

Notes

PREFACE

1. C. H. Schenck, S. R. Bundlie, M. G. Ettinger, M. W. Mahowald, "Chronic Behavioral Disorders of Human REM Sleep: A New Category of Parasomnia," *Sleep* 9, no. 2 (1986): 293–308.

2. C. H. Schenck, S. R. Bundlie, A. L. Patterson, M. W. Mahowald, "Rapid Eye Movement Sleep Behavior Disorder: A Treatable Parasomnia Affecting Older Adults," *Journal of the American Medical Association* 257 (1987): 1786–89.

3. C. H. Schenck, M. W. Mahowald, "REM Sleep Behavior Disorder: Clinical, Developmental, and Neuroscience Perspectives 16 Years After Its Formal Identification in *Sleep*," *Sleep* 25 (2002): 120–38.

4. C. H. Schenck, D. M. Milner, T. D. Hurwitz, S. R. Bundlie, M. W. Mahowald, "A Polysomnographic and Clinical Report on Sleep-Related Injury in 100 Adult Patients," *American Journal of Psychiatry* 146 (1989): 1166–73.

5. C. H. Schenck, T. D. Hurwitz, S. R. Bundlie, M. W. Mahowald, "Sleep-Related Eating Disorders: Polysomnographic Correlates of a Heterogeneous Syndrome Distinct from Daytime Eating Disorders," *Sleep* 14 (1991): 419–31.

6. M. W. Mahowald, S. R. Bundlie, T. D. Hurwitz, C. H. Schenck, "Sleep Violence— Forensic Implications: Polygraphic and Video Documentation," *Journal of Forensic Sciences* 35 (1990): 413–32.

7. M. W. Mahowald, C. H. Schenck, M. Goldner, V. Bachelder, M. Cramer-Bornemann, "Parasomnia Pseudo-Suicide." *Journal of Forensic Sciences* 48 (2003): 1158–62.

8. American Academy of Sleep Medicine. *International Classification of Sleep Disorders,* 2nd ed.: Diagnostic and coding manual. Westchester, Ill.: American Academy of Sleep Medicine, 2005.

INTRODUCTION

1. *The World Factbook,* CIA, 2005.

CHAPTER ONE

1. M. A. Carskadon, W. C. Dement, "Normal Human Sleep: An Overview." *Principles and Practice of Sleep Medicine,* 4th Ed., Elsevier Saunders, 2005.
2. "National Sleep Foundation 2004 *Sleep in America* Poll Highlights," National Sleep Foundation, 2004.
3. "Adolescent Sleep Needs and Patterns," Research Report and Resource Guide, National Sleep Foundation, 2000.
4. A. I. Pack, A. M. Pack, E. Rodgman, A. Cucchiara, D. F. Dinges, C. W. Schwab. "Characteristics of Crashes Attributed to the Driver Having Fallen Asleep." *Accident Analysis and Prevention* 27, no. 6 (1995): 769–75.
5. "*Sleep in America* Poll," National Sleep Foundation, 2002.
6. H. C. Hsu, M. H. Lin, "Exploring Quality of Sleep and Its Related Factors Among Menopausal Women," *Journal of Nursing Research* 13 no. 2 (2005): 153–64.
7. "Sleep in America Poll," National Sleep Foundation, 2003.
8. "Brain Basics: Understanding Sleep," National Institute of Neurological Disorders and Stroke, Pub. No. 04-3440-c, updated in 2005.
9. E. L. Lipizzi, B. P. Leavitt, D. B. Killgore, G. H. Kamimori, W. D. Killgore, "Decision Making Capabilities Decline with Increasing Duration of Wakefulness," *Sleep* 29 (2006): A131.
10. L. M. Day, C. J. Li, D. B. Killgore, G. H. Kamimori, W. D. Killgore, "Emotional Intelligence Moderates the Effect of Sleep Deprivation on Moral Reasoning," *Sleep* 29 (2006): A135.

CHAPTER TWO

1. J. D. Venable, "A Brief Biography of Thomas Alva Edison," The Charles Edison Fund.
2. M. Partinen, C. Hublin, "Epidemiology of Sleep Disorders," *Principles and Practice of Sleep Medicine,* 4th ed., Elsevier Saunders, 2005.

3. M. Partinen, et al., "Complaints of Insomnia in Different Occupations," *Scandinavian Journal of Work, Environment & Health* 10 (6 spec. no.; Dec. 1984): 467–69.

4. "Tunes Blared at Neighbor for Years 'Not Too Loud,'" *The Japan Times,* June 28, 2005.

5. M. M. Ohayon, "Severe Hot Flashes Are Associated with Chronic Insomnia, *Archives of Internal Medicine* 166, no. 12 (2006): 1262–68.

6. "New Study Pulls the Blanket Off America's Best and Worst Cities for Sleep," press release from Sperling's Best Places, www.bestplaces.net, 2004.

7. C. H. Schenck, M. W. Mahowald. "Long-Term, Nightly Benzodiazepine Treatment of Injurious Parasomnias and Other Disorders of Disrupted Nocturnal Sleep in 170 Adults," *American Journal of Medicine* 100, no. 3 (1996): 333–37.

8. C. H. Schenck, M. W. Mahowald, R. L. Sack, "Assessment and Management of Insomnia," *Journal of the American Medical Association* 289 (2003): 2475–79.

9. D. J. Buysse (ed., sect. 9), "Insomnia," *Principles and Practice of Sleep Medicine,* 4th ed., Elsevier Saunders, 2005.

10. M. O. Summers, M. I. Crisostomo, E. J. Stepanski, "Recent Developments in the Classification, Evaluation, and Treatment of Insomnia." *Chest* 130 (2006): 276–286.

CHAPTER THREE

1. F. J. Nieto, et al., "Association of Sleep-Disordered Breathing, Sleep Apnea, and Hypertension in a Large Community-Based Study; Sleep Heart Health Study," *JAMA* 283, no. 14 (2000): 1829–36.

2. S. Mazza, et al., "Most Obstructive Sleep Apnoea Patients Exhibit Vigilance and Attention Deficits on an Extended Battery of Tests," *European Respiratory Journal* 25 no. 1 (2005): 75–80.

3. L. J. Findley, M. E. Unverzagt, P. M. Suratt, "Automobile Accidents Involving Patients with Obstructive Sleep Apnea," *American Review of Respiratory Disease* 138, no. 2 (1988): 337–40.

4. M. Littner, et al., Standards of Practice Committee of the American Academy of Sleep Medicine, "Practice Parameters for the Use of Auto-Titrating Continuous Positive Airway Pressure Devices for Titrating Pressures and Treating Adult Patients with Obstructive Sleep Apnea Syndrome, *Sleep* 25 (2002): 143–end.

5. R. B. Berry, et al., An American Academy of Sleep Medicine Review, "The Use of Auto-Titrating Continuous Positive Airway Pressure for Treatment of Adult Obstructive Sleep Apnea, *Sleep* 25 (2002): 148–end.

6. C. A. Kushida, et al., An American Academy of Sleep Medicine Report, "Practice Parameters for the Use of Continuous and Bilevel Positive Airway Pressure Devices to Treat Adult Patients with Sleep-Related Breathing Disorders, *Sleep* 29, no. 3 (2006): 375–80.

7. P. Gay, et al., A Review by the Positive Airway Pressure Task Force of the Standards of Practice Committee of the American Academy of Sleep Medicine, "Evaluation of

Positive Airway Pressure Treatment for Sleep Related Breathing Disorders in Adults, *Sleep* 29, no. 3 (2006): 381–401.

8. C. A. Kushida, et al. An American Academy of Sleep Medicine Report, "Practice Parameters for the Treatment of Snoring and Obstructive Sleep Apnea with Oral Appliances: An Update for 2005, *Sleep* 29, no. 2 (2006): 240–43.

9. K. A. Ferguson, et al., "Oral Appliances for Snoring and Obstructive Sleep Apnea: A Review," *Sleep* 29, no. 2 (2006): 244–62.

10. K. K. Ki, "Surgical Therapy for Adult Obstructive Sleep Apnea," *Sleep Medicine Reviews* 9 (2005): 201–9.

11. J. Baron, D. Auckley, "Gunshot Wound to the Head: An Unusual Complication of Sleep Apnea and Bilevel Positive Airway Pressure," Sleep and Breathing 8, no. 3 (2004): 161–64.

12. O. Lateef, J. Wyatt, R. Cartwright, "A Case of Violent Non-REM Parasomnias That Resolved with Treatment of Obstructive Sleep Apnea," *Chest* 128 (2005): 461S.

CHAPTER FOUR

1. J. Montplaisir, R. P. Allen, A. S. Walters, L. Ferini-Strambi, "Restless Legs Syndrome and Periodic Limb Movements During Sleep," *Principles and Practice of Sleep Medicine*, 4th ed., Elsevier Saunders, 2005.

2. D. L. Pichietti, A. S. Walters, "Restless Legs Syndrome and Periodic Limb Movement Disorder in Children and Adolescents; Comorbidity with Attention-Deficit Hyperactivity Disorder," *Child and Adolescent Psychiatric Clinics of North America* 5 (1996): 729–40.

3. D. L. Pichietti, S. J. England, A. S. Walters, et al. "Periodic Limb Movement Disorder and Restless Legs Syndrome in Children with Attention-Deficit Hyperactivity Disorder." *Journal of Child Neurology* 13 (1998): 588–94.

4. D. L. Picchietti, D. J. Underwood, W. A. Farris, et al. "Further Studies on Periodic Limb Movement Disorder and Restless Legs Syndrome in Children with Attention-Deficit Hyperactivity Disorder." *Movement Disorders* 14 (1999): 1000–1007.

5. S. S. Rajaram, A. S. Walters, S. J. England, et al. "Some Children with Growing Pains May Actually Have Restless Legs Syndrome." *Sleep* 27 (2004): 767–73.

6. R. Ferri, B. Lanuzza, et al., "A Single Question for the Rapid Screening of Restless Legs Syndrome in the Neurological Clinical Practice," *Sleep* 29 (2006): A279.

7. R. P. Allen, D. Pichietti, W. A. Hening, C. Trenkwalder, A. S. Walters, and J. Montplaisir, "Restless Legs Syndrome: Diagnostic Criteria, Special Considerations, and Epidemiology; a Report from the Restless Legs Syndrome Diagnosis and Epidemiology Workshop at the National Institutes of Health." *Sleep Medicine* 4, no. 2 (2003): 101–19.

8. A. S. Walters, J. Winkelmann, C. Trenkwalder, et al., "Long-term Follow-up on Restless Legs Syndrome Patients Treated with Opioids." *Movement Disorders* 16 (2001): 1105–9.

9. A. H. Friedlander, M. E. Mahler, J. A. Yagiela, "Restless Legs Syndrome: Manifestations, Treatment, and Dental Implications," *Journal of the American Dental Association* 137, no. 6 (2006): 755–61.

10. C. H. Schenck, "Restless Legs Syndrome and Periodic Limb Movements of Sleep: Global Therapeutic Considerations" (guest editorial), *Sleep Medicine Reviews* 6 (2002): 247–51.

11. B. Hogl, S. Kiechl, J. Willeit, et al., "Restless Legs Syndrome: A Community-Based Study of Prevalence, Severity, and Risk Factors." *Neurology* 64 (2005): 1920-24.

CHAPTER FIVE

1. E. Morrish, et al., "Factors Associated with a Delay in the Diagnosis of Narcolepsy," *Sleep Medicine* 5 no. 2 (2004): 37–41.

2. C. H. Schenck, M. W. Mahowald, "Motor Dyscontrol in Narcolepsy: Rapid-Eye-Movement (REM) Sleep Without Atonia and REM Sleep Behavior Disorder," *Annals of Neurology* 32 (1992): 3–10.

3. S. Nightingale, et al., "The Association Between Narcolepsy and REM Behavior Disorder (RBD), *Sleep Medicine* 6 (2005): 253–58.

4. S. Marelli, M. Fantini, P. Busek, et al., "Association Between REM Sleep Behavior Disorder and Narcolepsy With and Without Cataplexy. *Sleep* 29 (2006): A229–A230.

5. National Institute of Neurological Disorders and Stroke, "Narcolepsy Fact Sheet," pub. no. 03-1637.

6. C. Guilleminault, S. Fromherz, "Narcolepsy: Diagnosis and Management," *Principles and Practice of Sleep Medicine,* 4th ed., Elsevier Saunders, 2005.

7. L. Lin, et al., "The Sleep Disorder Canine Narcolepsy Is Caused by a Mutation in the Hypocretin (Orexin) Receptor 2 Gene," *Cell* 98, no. 3 (1999): 365–76.

8. C. L. Bassetti, R. Pelayo, C. Guilleminault, "Idiopathetic Hypersomnia," *Principles and Practice of Sleep Medicine,* 4th ed., Elsevier Saunders, 2005.

9. I. Arnulf, J. M. Zeitzer, J. File, N. Farber, E. Mignot, "Kleine-Levin Syndrome: A Systematic Review of 186 Cases in the Literature, *Brain* 128 (2005): 2763–76.

CHAPTER SIX

1. K. Mundey, S. Benloucif, K. Harsanyi, M. L. Dubocovich, P. C. Zee, "Phase-Dependent Treatment of Delayed Sleep Phase Syndrome with Melatonin," *Sleep* 28 (2005): 1271–78.

2. J. T. Doljansky, et al., "Working Under Daylight Intensity Lamp: An Occupational Risk for Developing Circadian Rhythm Sleep Disorder? *Chronobiology International* 22 (2005): 597–605.

3. M. Okawa, "Circadian Rhythm Disorders in Sleep-Waking and Body Temperature in Elderly Patients with Dementia and Their Treatment, *Sleep* 14 (1991): 478–85.

4. M. M. Mallis, C. W. DeRoshia, "Circadian Rhythms, Sleep, and Performance in Space," *Aviation, Space, and Environmental Medicine* 76 (2005; 6 suppl): B94–107.

5. K. E. Bloch, "Transient Short Free Running Circadian Rhythm in a Case of Aneurysm Near the Suprachiasmatic Nuclei," *Journal of Neurology, Neurosurgery, and Psychiatry* 76 (2005): 1178–80.

6. C. L. Drake, et al., "Shift Work Sleep Disorder: Prevalence and Consequences Beyond That of Symptomatic Day Workers," *Sleep* 27 (2004): 1453–62.

7. F. W. Turek (section 8 editor), "Disorders of Chronobiology," *Principles and Practice of Sleep Medicine,* 4th ed., Elsevier Saunders, 2005.

CHAPTER SEVEN

1. A. Bonkalo, "Impulsive Acts and Confusional States During Incomplete Arousal from Sleep: Criminological and Forensic Implications," *Psychiatric Quarterly* 48, no. 3 (1974): 400–409.

2. M. Ohayon, R. Priest, J. Zulley, S. Smirne, "The Place of Confusional Arousals in Sleep and Mental Disorders: Findings in a General Population Sample of 13,057 Subjects," *Journal of Nervous and Mental Diseases* 188 (2000): 340–48.

3. B. Roth, S. Nevsimalova, A. Rechtschaffen, "Hypersomnia with 'Sleep Drunkenness,'" *Archives of General Psychiatry* 26 (1972): 456–62.

4. M. W. Mahowald, M. A. Cramer-Bornemann, "NREM Sleep-Arousal Parasomnias," *Principles and Practice of Sleep Medicine,* 4th ed., Elsevier Saunders, 2005.

5. C. H. Schenck, M. W. Mahowald, "Treatment of Severe Morning Sleep Inertia (SI) with Bedtime Sustained-Release (SR) Methylphenidate, Bupropion-SR, or Other Activating Agents," *Sleep* 26 (2003 suppl.): A75–76.

CHAPTER EIGHT

1. S. Saul, "Some Sleeping Pill Users Range Far Beyond Bed," *The New York Times,* March 8, 2006.

2. T. D. Hurwitz, M. W. Mahowald, C. H. Schenck, J. L. Schluter, S. R. Bundlie, "A Retrospective Outcome Study and Review of Hypnosis as Treatment of Adults with Sleepwalking and Sleep Terror," *Journal of Nervous and Mental Diseases* 179 (1991): 228–33.

3. C. Guilleminault, et al., "Adult Chronic Sleepwalking and Its Treatment Based on Polysomnography," *Brain* 128 (2005): 1062–69.

4. R. Broughton, et al., "Homicidal Somnambulism: A Case Report," *Sleep* 17, no. 3 (1994): 253–64.

5. O. Koster, T. Yaqoob, "Sleepwalker, Age 15, Found Curled Up on Crane," *Daily Mail* (UK), July 6, 2005.

6. "Sleeper Airborne," *Herald Sun,* June 17, 2005.

7. M. W. Mahowald, C. H. Schenck, M. A. Cramer-Bornemann, "Sleep-Related Violence," *Current Neurology and Neuroscience Reports* 5 (2005): 153–58.

8. W. Thomas Russell, " 'Nocturnal Seizure' Blamed for Slaying," *Press Telegram of Long Beach,* June 14, 2004.

9. P. Robin, "A Killer Sleep Disorder," *Phoenix New Times,* November 19, 1998.

10. D. Cohen, *Pillars of Salt, Monuments of Grace: New England Crime Literature and the Origins of American Popular Culture, 1674–1860* (Oxford University Press, 1993).

CHAPTER NINE

1. D. S. Rosenfeld, A. J. Elhajjar, "Sleepsex: A Variant of Sleepwalking," *Archives of Sexual Behavior* 27 (1998): 269–78.

2. C. Guilleminault, A. Moscovitch, K. Yuen, D. Poyares, "Atypical Sexual Behavior During Sleep," *Psychosomatic Medicine* 64 (2002): 328–36.

3. C. M. Shapiro, N. N. Trajanovic, J. P. Fedoroff, "Sexsomnia—A New Parasomnia?" *Canadian Journal of Psychiatry* 48 (2003): 311–17.

4. M. A. Mangan, "A Phenomenology of Problematic Sexual Behavior Occurring in Sleep," *Archives of Sexual Behavior* 33 (2004): 287–93.

5. N. Trajanovic, M. Mangan, C. M. Shapiro, "Sexual Behavior in Sleep: An Internet Survey," *Sleep* 29 (2006, suppl): A270.

6. "Sleeping Woman 'Seduced Strangers,' " *BBC News Magazine,* October 14, 2004.

7. R. Morgan, "Are You Raping Your Wife in Your Sleep?" *Details* 160, April 2006.

8. L. Redmond, "Probation in Chelmsford 'Sexsomnia' Case," *Lowell Sun,* May 19, 2005.

9. A. Dalton, "Sleepstalking," *Legal Affairs,* May/June 2005.

10. "Sleepwalking Suspect Convicted," Associated Press, August 6, 2005.

11. "Man Acquitted of Rape: Says He Was Asleep," Reuters, Oslo, December 10, 2004.

12. A. Farley, "Lust-Crazed Charity Worker Escapes Jail Term," Press Association News, UK, September 23, 2004.

13. S. Clough, "Landlord Gets Seven Years for Raping Sleepwalker," *The Daily Telegraph* (UK), October 25, 2001.

14. M. Mangan, *Sleepsex: Uncovered* (Philadelphia: Xlibris, 2001), 48, 59.

CHAPTER TEN

1. C. Schenck, M. Mahowald, "On the Reported Association of Psychopathology with Sleep Terrors in Adults, *Sleep* 23 (2000): 448–49.

2. C. H. Schenck, D. M. Milner, T. D. Hurwitz, S. R. Bundlie, M. W. Mahowald, "A Polysomnographic and Clinical Report on Sleep-Related Injury in 100 Adult Patients. *American Journal of Psychiatry* 146 (1989): 1166–73.

3. *Sleep Runners: The Stories Behind Everyday Parasomnias,* one-hour documentary (St. Paul, Minn.: Slow-Wave Films, LLC, 2004), www.sleeprunners.com.

4. C. Guilleminault, et al., "Sleepwalking and Sleep Terrors in Prepubertal Children: What Triggers Them?" *Pediatrics* 111, no. 1 (2003): e17–25.

5. A. Kales, C. Soldatos, E. Bixler, et al., "Hereditary Factors in Sleepwalking and Night Terrors, *British Journal of Psychiatry* 137 (1980): 111–18.

6. C. H. Schenck, M. W. Mahowald, "Two Cases of Premenstrual Sleep Terrors and Injurious Sleep-walking," *Journal of Psychosomatic Obstetrics and Gynaecology* 16 (1995): 79–84.

7. C. H. Schenck, *Paradox Lost: Midnight in the Battleground of Sleep and Dreams* (Minneapolis, Minn.: Extreme-Nights, LLC, 2005), www.parasomnias-rbd.com.

8. U. Isik, O. F. D'Cruz, "Cluster Headaches Simulating Parasomnias," *Pediatric Neurology* 27 (2002): 227–29.

9. M. W. Mahowald, G. M. Rosen, "Parasomnias in Children," *Pediatrician* 17, no. 1 (1990): 21–31.

CHAPTER ELEVEN

1. G. Goode, "Sleep Paralysis," *Archives Neurol* 6 (1962): 228–34.

2. M. Dahlitz, J. D. Parkes, "Sleep Paralysis," *Lancet* 341 (1993): 406–7.

3. T. Takeuchi, et al., "Isolated Sleep Paralysis Elicited by Sleep Interruption," *Sleep* 15, no. 3 (1992): 217–25.

4. R. J. McNally, et al., "Psychophysiological Responding During Script-Driven Imagery in People Reporting Abduction by Space Aliens," *Psychological Science* 15, no. 7 (2004): 493–97.

5. K. Fukuda, A. Miyasita, M. Inugami, K. Ishihara, "High Prevalence of Isolated Sleep Paralysis: Kanashibari Phenomenon in Japan." *Sleep* 10 (1987): 279–86.

6. H. Arikawa, et al., "The Structure and Correlates of Kanashibari," *Journal of Psychology* 133 (1999): 369–75.

7. D. J. Hufford, *The Terror That Comes in the Night: An Experience-Centered Study of Supernatural Assault Traditions* (Philadelphia: University of Pennsylvania Press, 1982).

CHAPTER TWELVE

1. J. Whyte, N. Kavey, "Somnambulistic Eating: A Report of Three Cases," *International Journal of Eating Disorders* 9 (1990): 577–81.

2. C. H. Schenck, T. D. Hurwitz, K. A. O'Connor, M. W. Mahowald, "Additional Categories of Sleep-Related Eating Disorders and the Current Status of Treatment, *Sleep* 16 (1993): 457–66.

3. J. Winkelman, "Clinical and Polysomnographic Features of Sleep-Related Eating Disorder, *Journal of Clinical Psychiatry* 58 (1998): 14–19.

4. K. C. Allison, A. L. Stunkard, S. L. Thier. *Overcoming Night Eating Syndrome.* Oakland, CA: New Harbingers Publications, Inc., 2004.

5. J. W. Winkleman, D. B. Herzog, M. Fava, "The Prevalence of Sleep-Related Eating Disorder in Psychiatric and Non-Psychiatric Populations," *Psychological Medicine* 29 (1999): 1461–66.

6. V. Paquet, et al., "Sleep-Related Eating Disorder Induced by Olanzapine," *Journal of Clinical Psychiatry* 63 (2002): 597.

7. M. L. Lu, W. W. Shen, "Sleep-Related Eating Disorder Induced by Risperidone," *Jounal of Clinical Psychiatry* 65 (2004): 273–74.

8. T. I. Morgenthaler, M. H. Silber, "Amnesic Sleep-Related Eating Disorder Associated with Zolpidem," *Sleep Medicine* 3 (2002): 323–27.

9. C. H. Schenck, "Zolpidem-Induced Sleep-Related Eating Disorder (SRED) in 19 Patients," *Sleep* 28 (2005; suppl): A259.

10. P. Roper, "Bulimia While Sleepwalking: A Rebuttal for Same Automatism?"; *Lancet* 2, no. 8666 (1989): 796.

11. S. G. Boodman, "Hungry in the Dark: Some Sleepeaters Don't Wake Up for Their Strange Nighttime Binges," *Washington Post,* HE01, September 7, 2004.

12. C. H. Schenck, M. W. Mahowald, "Topiramate Therapy of Sleep-Related Eating Disorder (SRED)." *Sleep* 29 (Suppl 2006): A268.

13. J. W. Winkelman, "Efficacy and Tolerability of Topiramate in the Treatment of Sleep-Related Eating Disorder: An Open-label, Retrospective Case Series. *Journal of Clinical Psychiatry* 2006 (in press).

14. G. Keillor, "The Old Scout: To Conquer Sleep Eating, First Face Up to the Truth," *Minneapolis-St. Paul Star Tribune,* February 19, 2006.

CHAPTER THIRTEEN

1. M. W. Mahowald, W. C. Dement, C. H. Schenck, "REM Sleep Parasomnias," in M. H. Kryger, T. Roth, eds. *Principles and Practice of Sleep Medicine,* 4th ed., Philadelphia: Elsevier Saunders, 2005: 897–916.

2. M. L. Fantini, A. Corona, S. Clerici, L. Ferini-Strambi, "Aggressive Dream Content Without Daytime Aggressiveness in REM Sleep Behavior Disorder, *Neurology* 65 (2005): 1010–15.

3. E. Olson, B. Boeve, M. Silber, "Rapid Eye Movement Sleep Behavior Disorder: Demographic, Clinical, and Laboratory Findings in 93 Cases," *Brain* 123 (2000): 331–39.

4. C. Schenck, S. Bundlie, M. Mahowald, "Delayed Emergence of a Parkinsonian Disorder in 38 Percent of 29 Older Men Initially Diagnosed with Idiopathic Rapid Eye Movement Sleep Behavior Disorder," *Neurology* 46 (1996): 388–93.

5. C. H. Schenck, et al., "REM Behavior Disorder (RBD): Delayed Emergence of Parkinsonism and/or Dementia in 65 Percent of Older Men Initially Diagnosed with Idiopathic RBD, *Sleep* 26 (2003 suppl.): A316.

6. A. Iranzo, et al., "Rapid Eye Movement Sleep Behavior Disorder as an Early Marker for a Neurodegenerative Disorder: A Descriptive Study, *Lancet Neurology* 5 (2006): 572–77.

7. J. F. Gagnon, R. B. Postuma, S. Mazza, J. Doyon, J. Montplaisir, "Rapid-Eye-Movement Sleep Behaviour Disorder and Neurodegenerative Diseases," *Lancet Neurology* 5 (2006): 424–32.

8. S. B. Yeh, C. H. Schenck, "A Case of Marital Discord and Secondary Depression with Attempted Suicide Resulting from REM Sleep Behavior Disorder in a 35-Year-Old Woman, *Sleep Medicine* 5 (2004): 151–54.

9. C. Schenck, J. Boyd, M. Mahowald, "A Parasomnia Overlap Disorder Involving Sleepwalking, Sleep Terrors, and REM Sleep Behavior Disorder in 33 Polysomnographically Confirmed Cases, *Sleep* 20 (1997): 972–81.

10. J. W. Winkelman, L. James, "Serotonergic Antidepressants Are Associated with REM Sleep Without Atonia," *Sleep* 27, no. 2 (2004): 317–21.

CHAPTER FOURTEEN

1. C. Schenck, D. Milner, T. Hurwitz, S. Bundlie, M. Mahowald, "Dissociative Disorders Presenting as Somnambulism: Video and Clinical Documentation (8 Cases)," *Dissociation* 2 (1989): 194–204.

2. E. Rice, C. Fisher, "Fugue States in Sleep and Wakefulness: A Psychophysiological Study," *Journal of Nervous and Mental Diseases* 163 (1976): 79–87.

3. J. Fleming, "Dissociative Episodes Presenting as Somnambulism: A Case Report," *Sleep Research* 16 (1987): 263.

4. P. Coons, E. Bowman, "Dissociation and Eating," *American Journal of Psychiatry* 150 (1993): 171–72.

5. M. Y. Agargun, et al., "Characteristics of Patients with Nocturnal Dissociative Disorders," *Sleep and Hypnosis* 3 (2001): 131–34.

6. M. Mahowald, C. Schenck, "Nocturnal Dissociation: Awake? Asleep? Both? Or Neither?" *Sleep and Hypnosis* 3 (2001): 129–30.

CHAPTER FIFTEEN

1. R. N. Reimão, A. B. Lefévre, "Prevalence of Sleep-Talking in Childhood," *Brain and Development* 2 (1980): 353–57.

2. "Report: Sleep 'Divorce' Counts," Reuters, March 27, 2006.

3. "National Briefs," *Lawrence Journal-World,* April 14, 2001.

4. *Godfrey v. State,* 358 S.E. 2d 264 (Ga. Ct. App. 1987).

5. J. A. Pareja, et al., "A First Case of Progressive Supranuclear Palsy and Pre-Clinical REM Sleep Behavior Disorder Presenting as Inhibition of Speech During Wakefulness and Somniloquy with Phasic Muscle Twitching During REM Sleep," *Neurologia* 11 (1996): 304–6.

6. N. Tachibana, et al., "REM Sleep Motor Dysfunction in Multiple System Atrophy: With Special Emphasis on Sleep Talk as Its Early Clinical Manifestation, *Journal of Neurology, Neurosurgery, and Psychiatry* 63 (1997): 678–81.

7. J. A. Pareja, et al., "Native Language Shifts Across Sleep-Wake States in Bilingual Sleeptalkers, *Sleep* 22 (1999): 243–47.

8. N. Lukianowicz, "Auditory Hallucinations in Polyglot Subjects," *Psychiatria et Neurologia* 143 (1962): 274–94.

CHAPTER SIXTEEN

1. M. R. Pressman, et al., "Night Terrors in an Adult Precipitated by Sleep Apnea," *Sleep* 18 (1995): 773–75.

2. A. Iranzo, J. Santamaria, "Severe Obstructive Sleep Apnea/Hypopnea Mimicking REM Sleep Behavior Disorder," *Sleep* 28 (2005): 203–6.

3. R. P. Millman, G. J. Kipp, M. A. Carskadon, "Sleepwalking Precipitated by Treatment of Sleep Apnea with Nasal CPAP," *Chest* 99 (1991): 750–51.

4. T. T. Sjoholm, et al., "Sleep Bruxism in Patients with Sleep-Disordered Breathing," *Archives of Oral Biology* 45 (2000): 889–96.

5. O. Devinsky, et al., "Epilepsy and Sleep Apnea Syndrome," *Neurology* 44 (1994): 2060–64.

6. B. A. Malow, et al., "Epilepsy and Sleep Apnea Syndrome," *Sleep Medicine* 4, no. 6 (2003): 509–15.

7. J. L. Goodwin, et al., "Parasomnias and Sleep Disordered Breathing in Caucasian and Hispanic children—the Tucson Children's Assessment of Sleep Apnea Study," *BMC Medicine* 2 (2004): 14.

8. C. Guilleminault, et al., "Sleepwalking and Sleep Terrors in Prepubertal Children: What Triggers Them?" *Pediatrics* 111 (2003): 17–25.

9. J. Owens, et al., "Incidence of Parasomnias in Children with Obstructive Sleep Apnea," *Sleep* 20 (1997): 1193–96.

10. M. M. Ohayon, et al., "Violent Behavior During Sleep," *Journal of Clinical Psychiatry* 58 (1997): 369–76.

CHAPTER SEVENTEEN

1. N. N. Trajanovic, C. M. Shapiro, "Sleep-Laughing (Somnorismus)," *Sleep* 28 (2005): A260.
2. J. M. Pearce, "Clinical Features of the Exploding Head Syndrome," *Journal of Neurology, Neurosurgery, and Psychiatry* 52 (1989): 907–10.
3. C. Sachs, E. Svanborg, "The Exploding Head Syndrome: Polysomnographic Recordings and Therapeutic Suggestions," *Sleep* 14 (1991): 263–66.
4. R. Vetrugno, et al., "Catathrenia (Nocturnal Groaning): A New Type of Parasomnia," *Neurology* 56 (2001): 681–83.
5. J. Iriarte, et al., "Continuous Positive Airway Pressure as Treatment for Catathrenia (Nocturnal Groaning)," *Neurology* 66 (2006): 609.
6. C. H. Schenck, et al., "Prominent Eye Movements During NREM Sleep and REM Sleep Behavior Disorder Associates with Fluoxetine Treatment of Depression and Obsessive-Compulsive Disorder," *Sleep* 15 (1992): 226–35.
7. J. L. Davis, D. C. Wright, "Exposure Relaxation, and Rescripting Treatment for Trauma-Related Nightmares," *Journal of Trauma Dissociation* 7, no. 1 (2005): 5–18.
8. T. A. Nielsen, A. Zadra. "Nightmares and Other Common Dream Disturbances." In M. H. Kryger, T. Roth, W. C. Dement, eds., *Principles and Practice of Sleep Medicine,* 4th ed. Philadelphia: Elsevier Saunders, 926–35.
9. M. A. Raskind, et al., "Reduction of Nightmares and Other PTSD Symptoms in Combat Veterans by Prazosin: A Placebo-Controlled Study," *American Journal of Psychiatry* 160 (2003): 371–73.
10. S. Van Liempt, E. Vermetten, E. Geuze, H. Westenberg. "Pharmacotherapeutic Treatment of Nightmares and Insomnia in Posttraumatic Stress Disorder: An Overview of the Literature." *Annals of the New York Academy of Sciences* 1071 (2006): 502–7.

CHAPTER EIGHTEEN

1. A. S. Baran, A. C. Richert, R. Goldberg, J. M. Fry, Posttraumatic Stress Disorder in the Spouse of a Patient with Sleep Terrors, *Sleep Medicine* 4 (2003): 73–75.

General References on Sleep Disorders and Parasomnias

American Academy of Sleep Medicine. *International Classification of Sleep Disorders,* 2nd ed. Westchester, Ill., 2005.

Dement, W. C., C. Vaughan. *The Promise of Sleep.* Delacourt Press, 1999.

Hauri, P., S. Linde. *No More Sleepless Nights.* New York: Wiley & Sons, 1996.

Kryger, M. H., T. Roth, W. C. Dement, eds. *Principles and Practice of Sleep Medicine,* 4th ed. Philadelphia: Elsevier Saunders, 2005.

Schenck, C. H. *Paradox Lost: Midnight in the Battleground of Sleep and Dreams.* Minneapolis, Minn.: Extreme-Nights, LLC, www.parasomnias-rbd.com.

Schenck, C. H., B. Dehler. *Sleep Runners: The Stories Behind Everyday Parasomnias.* St. Paul, Minn: Slow-Wave Films, LLC, www.sleeprunners.com.

Schenck, C. H., M. W. Mahowald, "Parasomnias from a Woman's Health Perspective." In H. P. Attarian, ed. *Sleep Disorders for Gynecologists and Primary Care Physicians.* Totowa, N.J.: Humana Press, 2006.

Contact the National Sleep Foundation (www.sleepfoundation.org) for additional information about sleep disorders and to find the names of accredited sleep disorders centers in your area. Please note that some sleep disorders centers are accredited only for sleep apnea testing. Also, not all sleep disorders centers are experienced in evaluating and treating parasomnias, so please specify if a parasomnia is your primary problem needing evaluation.

14